Filter Dust Collectors

Other McGraw-Hill Chemical Engineering Books of Interest

BRUNNER • *Hazardous Waste Incineration*

CHOPEY • *Handbook of Chemical Engineering Calculations*

CHOPEY • *Environmental Engineering in the Process Plant*

COOK, DUMONT • *Process Drying Practice*

DEAN • *Lange's Handbook of Chemistry*

DILLON • *Materials Selection for the Chemical Process Industries*

FREEMAN • *Hazardous Waste Minimization*

FREEMAN • *Standard Handbook of Hazardous Waste Treatment and Disposal*

KISTER • *Distillation Design*

KISTER • *Distillation Operation*

KOLLURU • *Environmental Strategies Handbook*

LEVIN, GEALT • *Biotreatment of Industrial and Hazardous Waste*

MCGEE • *Molecular Engineering*

MANSFIELD • *Engineering Design for Process Facilities*

MILLER • *Flow Measurement Engineering Handbook*

PALLUZI • *Pilot Plant and Laboratory Safety*

PALLUZI • *Pilot Plant Design, Construction, and Operation*

PERRY, GREEN • *Perry's Chemical Engineers' Handbook*

POWER • *Steam Jet Ejectors for the Process Industries*

REID ET AL. • *Properties of Gases and Liquids*

REIST • *Aerosol Science and Technology*

SANDLER, LUCKIEWICZ • *Practical Process Engineering*

SATTERFIELD • *Heterogenous Catalysis in Industrial Practice*

SHINSKEY • *Feedback Controllers for the Process Industries*

SHINSKEY • *Process Control Systems*

SHUGAR, BALLINGER • *The Chemical Technicians' Ready Reference Handbook*

SHUGAR, DEAN • *The Chemist's Ready Reference Handbook*

SMITH, VAN LAAN • *Piping and Pipe Support Systems*

STOCK • *AI in Process Control*

TATTERSON • *Fluid Mixing and Gas Dispersion in Agitated Tanks*

TATTERSON • *Scaleup and Design of Industrial Mixing Processes*

WILLIG • *Environmental TQM*

YOKELL • *A Working Guide to Shell-and-Tube Heat Exchangers*

Filter Dust Collectors

Design and Application

Miles L. Croom
Engineering Manager, DCAD Company
Lexington, North Carolina

McGraw-Hill, Inc.
New York San Francisco Washington, D.C. Auckland Bogotá
Caracas Lisbon London Madrid Mexico City Milan
Montreal New Delhi San Juan Singapore
Sydney Tokyo Toronto

Library of Congress Cataloging-in-Publication Data

Croom, Miles L.
Filter dust collectors / Miles L. Croom.
p. cm.
Includes bibliographical references and index.
ISBN 0-07-014500-8 (alk. paper)
1. Air filters. I. Title.
TH7683.A3C76 1994
697.9′324—dc20 94-17463
CIP

1 2 3 4 5 6 7 8 9 0 DOC/DOC 9 0 9 8 7 6 5 4

ISBN 0-07-014500-8

The sponsoring editor for this book was Gail F. Nalven, the editing supervisor was David E. Fogarty, and the production supervisor was Suzanne W. Babeuf. It was set in Century Schoolbook by McGraw-Hill's Professional Book Group composition unit.

Printed and bound by R. R. Donnelley & Sons Company.

This book is printed on recycled, acid-free paper containing a minimum of 50% recycled, de-inked fiber.

I dedicate this book to the people who will further improve pollution control in the future, and to my wife Eileen for her understanding and support

Contents

Preface

"You've got to know when to push it; and when to back off!" CHUCK YEAGER

Industrial dust collectors need to be cost-effective and increasingly efficient in capturing fine dust particles to meet ever tighter emission limits. Filter dust collectors usually attain these goals better than any other method. Much has been written about dust collection theories, and various proprietary design advantages have been extolled. But there is an increasing need for objective guidelines on the overall technology and industrial application of filter dust collection system.

The goal of this book will be to present concise guides for developing and evaluating the optimum filter dust collector system for each application. These guides should be an especially useful reference when the project time is short and the emission limits are tight.

Filter dust collectors are being used on particle size ranges down to 0.3 micron, temperatures to 500°F (260°C), and at efficiencies of 99.99 percent and better. Ceramic and metal filter media are being developed and tested that will extend the application range to and above operating temperatures of 1000°F (540°C).

Combinations of functions in one enclosure are also being developed to neutralize sulfur dioxide, filter fine dust particles, and reduce nitrous oxides. This is being done successfully to meet emission limits on all three air pollutants at the optimum operating temperature level of around 850°F (455°C).*

Successful dust collection requires a complete system that will capture, deliver, collect, and discharge dust products efficiently and at a justifiable cost level. This book will provide practical guidelines for all facets of a complete and optimum filter dust collector system. The ultimate guide to follow in dust collection is what Chuck Yeager advised about flying, "You've got to know when to push it, and when to back off!"

Miles L. Croom

**Chemical Engineering*, Sept. 1993.

Chapter

1

Evaluating and Planning Collectors

"Investment in Quality never costs anything."

W. EDWARDS DEMING

The first priority in planning a collection system should be to determine the project's basic data as fully as possible. This should include specific characteristics of the dust that will be collected, those of the process or conveying airstream or gas, the permissible emission limits, and the plant location and site requirements. Then evaluation of pollution prevention or control plans can be made.

1.1 Basic Data

Before deciding what type of collector to use, gather all possible information immediately and obtain any further data as the project proceeds.

Dust or product

The following data should be collected about the dust or product to be filtered:

1. Particle size range and percentage or count in each size level, particularly the fines (see Fig. 1.1).
2. Bulk density of dust, both packed and aerated.
3. Process dust loading capacity required in grains per cubic foot (gr/ft^3) [milligrams per cubic meter (mg/m^3)] of airstream, or in pounds per hour (lb/h) [kilogram per hour (kg/h)] fed or captured

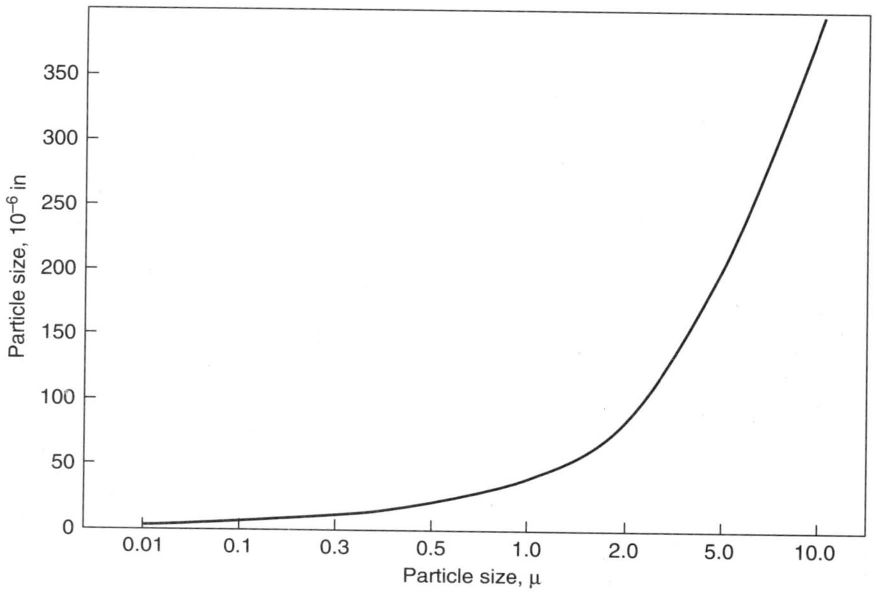

Figure 1.1 Particle size conversions. Micron particle size versus inch size (where 1 μ equals 3.937×10^{-5} in, and 10 μ equal 3.937×10^{-4} in).

by the system, or expressed as volume of product per hour. Both normal and maximum operating levels must be considered.

4. Weight or volume of dust to be stored or held in the collector, if needed.
5. Any special characteristics of the dust or airstream, such as temperature, corrosiveness or pH, abrasiveness, friability, moisture content, dew point, explosiveness, flammability, flowability, and tendency to agglomerate, fluidize or pack, be hygroscopic, or degrade.
6. Potential savings through improved yield, quality, recovery, or reuse in some other plant process rather than end-of-pipe collection and disposal.

Airstream or gas

The following data should be obtained about the airstream or gas to be filtered:

1. Volume required in cubic feet per minute (ft^3/min) [cubic meters per minute (m^3/min)] at both normal and maximum rates
2. Temperature of the airstream, dust, and outside atmosphere, again for all conditions

3. Whether insulation or heaters should be considered
4. Any special characteristics of the airstream
5. Potential energy recovery or other savings, such as reuse of the cleaned airstream for makeup air or combustion air supply
6. Pressure drop expected in the capture and delivery ductwork, in the collector, and in dampers or other accessories

Emission limits

1. Clarify allowable emission limits and conditions.
2. Determine what may be the best available control technology (BACT) for this application and what other solutions are feasible.
3. Determine the possibility of achieving zero emissions through some combination of process design or change, recycling, or reuse that might eliminate the collection or emission control need altogether.

Other application needs or possibilities

The following information may also be pertinent:

1. Wind or seismic load, special structural strength, or foundation design for the soil
2. Safety or insurance requirements or benefits for access, hazard control [explosion vents, sprinklers, fire suppression, NEMA (National Electrical Manufacturers Association) enclosure ratings, and sparkproof construction]
3. Materials of construction or protective coatings
4. Module or multiple baghouse construction, isolation dampers, bypass dampers, and ducts
5. Accessories, such as level controls, flow aids, control or instrumentation panels, motion switches, noise limiting controls, fan volume damper, heaters or insulation, feeder or discharge device, and others
6. Space or site limitations, height, area, or access
7. Expected hours of operation per year, or service life

1.2 Preliminary Evaluation

The search for the best pollution solution can proceed from the basic data to an evaluation of all reasonable alternatives and preparation of a preliminary system proposal. Look for the most cost-effective type of collection for the emission limits required. And consider new or improved technology. If a preliminary budget request is needed, it can be based on this early spadework.

Pollution control should be considered among the best early project designs, rather than being an afterthought. Too often the plant or project design is nearly full blown before emissions are considered. Then it may be too late to include any alternatives or process changes to eliminate or reduce pollutant discharge in the operation. As the old jazz song says, "I sent for you yesterday, and here you come today."

1.3 Source Reduction

An article in *Chemical Engineering—Environmental Engineering Supplement* makes the case very well[1] by reporting that approximately $2.6 billion was spent by the U.S. manufacturing industry in 1990 on air pollution control. But of this amount only 29 percent was spent on source reduction techniques. The other 71 percent went for end-of-pipe solutions by pollution control.

Zero discharge is the ultimate source reduction goal and may be achieved by a closed loop or other process recovery change; a change of fuels, raw materials, or neutralization treatment; maintenance or design improvements; reuse or conversion of the pollutant; and other possibilities. Next best would be reduction of the discharge to a level below the required limits. Failing that, pollution controls must collect or reduce the pollutant to meet emission limits. And the waste must still be disposed of in some acceptable manner.

Case Study A case in point is the conversion of dry dust, wastewater, and sludge discharge from a ceramic tile plant into fired low-quality tile or loose aggregate for road construction. All these waste components are gathered, dried if necessary, and fired in pressed form for tile or bulk form for aggregate. The firing temperature is high enough, above 1800°F (982°C), to change them into an igneous product that is inert.

There is no hazard from use of the inert product, there is zero discharge from the plant's end-of-pipe, there would be some yield gain or additional income from the sale of the tile or aggregate, and there need be no further investment in pollution control equipment. Some 30 years' accumulation of scrap tile can be ground up and added to the "output," since it is already fired and inert.

The process changes eliminate the continuing waste output, as well as the future responsibility for its disposal. This solution also provides for the cleanup of old settling ponds, plus the dust and scrap tile piles, which each presented a surface runoff pollution problem. An added benefit came when the idea won a $10,000 "Pollution Prevention Challenge Grant" from the state in 1989 for the innovative "pollution solution."[2]

1.4 Pollution Control

Past successful experience or preliminary tests and collector evaluation should point out the best control alternatives and their approxi-

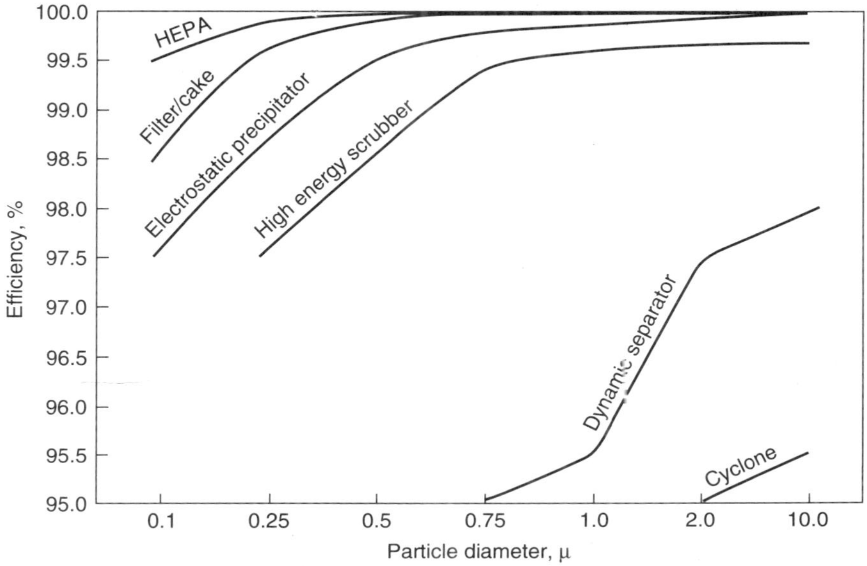

Figure 1.2 Collection efficiency versus particle size by type of collector.

mate cost ranges. Figure 1.2 highlights the efficiency versus particle size application ranges for different collector types. This can be used as a guide for the initial selection based on performance and minimum particle size required.

Preliminary selection can start with the specific dust particle size range and properties, as well as the airstream characteristics. Evaluation can be made by comparison of the projected performance, efficiency, space, energy, pressure drop, service temperature, emissions limits, and approximate cost for operation and investment. Realize that this is just a preliminary guide, since there are so many variables of design, materials, accessories, and specific application needs.

Therefore, unless the application is very similar to a previous job, each project will need to have specifications written and bids taken based on the preliminary plans, "because on the average, conditions are not average."[4] So a preliminary selection should be made as early as possible to help develop the final and best solution, as mentioned in Sec. 1.2. So when you're under the gun on a deadline or want to consider savings in project plans, develop a potential plan early!

One supplier's advertising asks the question, "Are your dust collecting techniques gathering dust?"[5] Keep this in mind as you prepare and evaluate plans. There may be newer developments whose time has come to be considered seriously. Chapter 12 suggests the inclusion in the bidding process of performance guarantees, acceptance testing, and the request to suppliers for recommended alternatives that may improve performance. This should be a catalyst in choosing the best solution.

1.5 Primary Collectors

Several decades ago these were the main collectors used, but now they usually serve to collect coarser particle sizes ahead of more efficient collectors. But when a heavy dust load of large particles, such as wood chips, needs to be separated, these will still need to be considered. A word of caution, nearly every type of collector needs an air seal discharge valve to prevent unwanted airflow from upsetting the effective collection operation (see Chap. 6).

Settling chambers

An enclosure that is larger enough in cross section than the conveying duct so that it reduces the airstream velocity below a dropout or settling rate for the desired dust particles is a settling chamber, or dropout box. The dust will settle to the bottom of the chamber for discharge, and the cleaned conveying air can exit through an outlet at a normal duct velocity.

Often an impingement or baffle plate is positioned just inside the inlet to help "knock down" the dust. The air velocity passing around the baffle and through the dropout section should be low enough that it will not reentrain, recapture, or pick up the desired dust particles. Because of their relatively low efficiency, these old favorites rarely make the cut nowadays.

However, for large particle sizes of around 40 microns (μ) and up, the settling chamber can serve as a primary collector. Products like sawdust, plastic pellets, feed and grain, chemicals, and food products in large sizes can be considered. The small target box collector at the discharge end of many pneumatic conveyors is an example of a still useful settling chamber. Figure 1.3 shows the cross section of a simple dropout box or settling chamber, using gravity separation.

Low pressure drops of 1 to 3 inches water gauge (in WG) can be achieved, initial costs are relatively low at \$0.50 to \$1.00 per ft^3/min, service temperature can be whatever the materials of construction will stand, but the collection efficiency generally will be in the 50 to 75 percent range. Sizing is based on reducing the flow velocity enough to permit settling of the desired dust. This will usually need to be

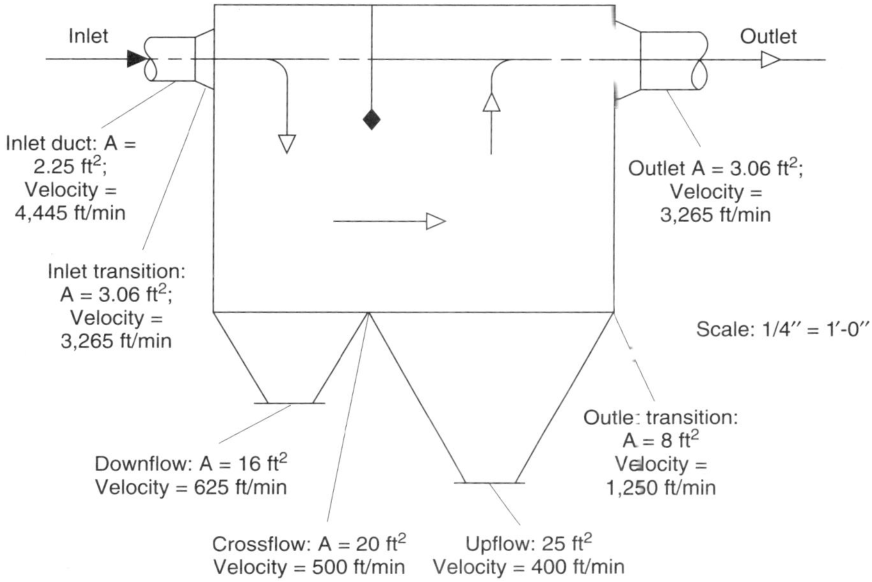

Figure 1.3 Settling chamber and gravity separation with cross-section area showing airstream velocity.

down in the 600- to 200-ft/min range. The flow velocity can be calculated by dividing the airstream volume by the flow cross-sectional area of the chamber. The large floor area or space required can become a problem on large flow volumes.[6]

A 10,000-ft^3/min volume should enter through a 5-ft^2 transition area to reduce the inlet flow velocity to around 2000 ft/min to minimize turbulence and pressure drop losses (see Chap. 8). Then in the dropout zone the flow cross-sectional area should be enlarged to around 50 ft^2 in order to slow the airstream velocity down to 200 ft/min, or the velocity needed for the application.

The inlet transition and settling chamber concept is a common design adaptation in other collectors for reducing the inlet velocity to 2000 ft/min or lower for the benefits of reduced turbulence and inlet pressure drop, and for dropout in the hopper. Usually some form of inlet baffle or diffuser plate will be included to help knock down the dust, reduce impact abrasion, and distribute the airflow uniformly. A further velocity reduction to around 200 ft/min should be considered for the succeeding flow passage into the collector section. Limiting the flow passage velocity between filters in a pulse-jet collector will minimize reentrainment of fine particles during filter cleaning cycles (see Chap. 2).

Cyclones

This is basically a cylinder with a horizontal tangential inlet, a top vertical outlet designed to produce a lower velocity than the inlet, and a tapered conical hopper bottom with an air lock dust discharge outlet valve. The incoming airflow is deflected by the tangential inlet and cylinder walls into a cyclone pattern, causing a rapid change in direction that separates particles from the airstream by centrifugal force and impact on the walls. Figure 1.4 depicts the cross section of a typical higher-efficiency cyclone collector design.[7]

Since the dust particles are denser than air, they are driven up against the walls by the cyclonic airflow. That impingement, friction against the walls, and the change in direction cause the particles to slow down much more than the airstream. So they drop out of the airstream and slide down into the hopper.

A low-pressure-drop cyclone can provide a collection efficiency up to 80 to 95 percent on particle sizes of 10 to 40 μ, respectively, at a pressure drop of around 2 to 3 in WG. For a 10,000-ft³/min (284.5 m³/min)

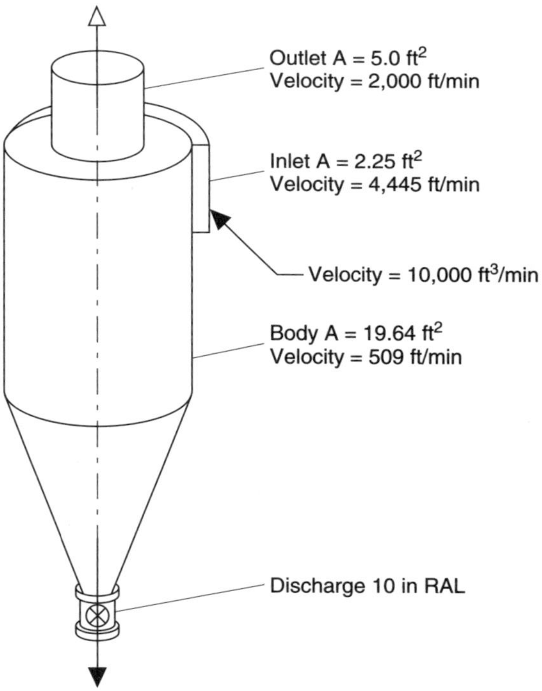

Figure 1.4 Cyclone collector operation and cross-section area showing centrifugal separation and airstream velocity.

flow volume, the diameter of the cyclone would be in the range of 42 to 68 in (16.5 to 26.8 cm). It would occupy a floor area of 9.62 to 25.2 ft^2 (0.9 to 2.35 m^2). An estimated ballpark cost would be around \$0.50 to \$0.80 per cubic feet per minute, or about a \$5000 to \$8000 initial cost for the collector and support structure only (1993 estimated cost levels).[3]

A high-efficiency cyclone can achieve around 90 to 97 percent efficiency, with a pressure drop of 3 to 5 in WG, on particle sizes of 5 to 10 μ. Higher-efficiency, smaller-diameter cyclones can be used on particles down to a size of 3 μ, but the pressure drop and the cost need to be considered. Three- to five-inch-diameter cyclones in multiple banks, or multiclones, have been commonly used on boiler and furnace exhausts at temperatures up to 800°F. The initial cost for the high-efficiency cyclones will run in the range of \$0.85 to \$1.75 per ft^3/min.

Manufacturers build cyclones to meet appropriate specifications in those ranges and will test or certify the performance. Cyclone investment and operating costs are relatively low if the emission limits can be met. Abrasive wear and corrosion may result in a short service life. Energy costs will become significant as the pressure drop increases and raises fan horsepower in proportion (see Fig. 1.5).

Figure 1.5 Cyclone and five reverse-air filter collectors, tangential inlet to filter collectors, five rotary air locks discharge to one pneumatic conveyor to cyclone collector, upper right, to sawdust storage silo for boiler fuel utilization and savings, and explosion vent panels in collector walls.

Cyclones can be used as a receiver for process dust collected by upstream collectors.

Elutriators

These collectors use the principle that "the rate at which particles settle in air is a function of particle size, shape, and density."[8] Reducing volume of flow and thus velocity in a vertical chamber applies the elutriation principle by gradually exhausting air volume at incremental heights as the dirty air rises. So larger particles drop out near the bottom and the smaller and lighter particles rise in the airstream and settle out only when the velocity is sufficiently reduced. The elutriator can thus be used as a particle size classifier by pulling off portions of the air volume and dust at appropriate incremental heights. Some elutriators can be adjusted in operation for the range of particle sizes separated by varying the air volume removed at selected heights.[8]

This is similar to the vertical separation that occurs in a filter collector as the dusty airstream rises and air volume is removed through the filters. Sophisticated settling chambers can be adjusted for particle size separation by using gradual exhaust of air volume. In tall shaker or reverse-air–cleaned filters used in power plants, and elsewhere, the bag length may reach 30 to 40 ft (9.15 to 12.2 m). The elutriation effect can cause some loss of the effective filter area by reentraining very small particles at the upper end of the bags which reduces airflow at the upper levels of the filter bag. This puts an operating limit on the bag length that may be used effectively.

Dynamic separators

Motor-driven impellers can speed up the velocity of the airstream driven into a cyclone-type separator to improve the collection efficiency. Other designs use independent or auxiliary airstreams to accelerate dust particles and thus increase the separation efficiency. The particle size collected can be as small as 1 μ.

The additional energy will increase costs, but efficiencies up to 99 percent are achieved. The separator impeller or fan can serve as the system fan in some cases. Obviously the abrasive wear will increase as the velocity increases, but floor space area is minimized. Costs should run from \$2.00 to \$3.00 per ft^3/min.

Electrostatic precipitators

The gas stream is passed through electrically charged grids of wires or plates, where the dust particles pick up a charge.[9] The airstream then passes through a second set of wires or plates with an opposite charge, which attract the dust particles. Periodically the collecting plates or wires are shaken or vibrated, causing the dust to fall off into a storage hopper, or modular sections can be closed off and the charge removed for periodic cleaning.[10]

Precipitators can collect particles as small as those in cigarette smoke, or about 0.1 μ. They have a collection efficiency of 98 to 99.9 percent and are used on large-volume airstreams at temperatures up to 1000°F (540°C). The initial cost may run from $1.50 to $4.00 per ft^3/min. Electrostatic precipitators are widely used on power plant and boiler fly ash separation at temperatures up to 1000°F.[11]

Wet scrubbers

These devices use water spray to collect dust, mist, or fumes from an airstream passing through the spray. Medium-energy types can collect up to 98 percent of the dust. High-energy venturi scrubbers can reach efficiencies up to 99.5 percent on particles as small as 0.5 μ. Appropriation costs run around $1.00 up to $2.00 per ft^3/min.

High-energy scrubbers require, as the name implies, a lot of energy for the associated high pressure drop. So the operating cost and the cost of disposal of collected dust must be considered in the cost analysis of this type of collector. The disposal of collected wet dust or sludge may become a costly step, involving wastewater treatment or a separation system. On the other hand some wet discharge materials may be directly returned, recycled, or reused in the process.

Gas absorbers

These devices use a reactive liquid or solution in a spray tower, packed column, or fiber cell collector to remove soluble or reactive gas contaminants from an airstream. They will remove pollutant fumes in sizes down to 0.001 μ from process exhaust. Solid absorbents such as lime or calcium carbonate are injected in powder form into the airstream to neutralize sulfur dioxide SO_2 in the airstream.

Gas adsorbers

These devices depend upon activated carbon, alumina, silica gel, or molecular sieves which collect contaminants by the physical process of gas molecules adhering to the surface of a solid. The adsorbent can be either regenerated or discarded after use. Particles collected are molecular in size, down to 0.001 μ.

Incinerators

These thermal devices are being used more frequently for industrial wastes or hazardous waste site cleanup, such as superfund projects, and they burn contaminants in the soil or waste materials either thermally or catalytically. This forms easily disposable products of combustion, such as carbon dioxide, water vapor, and nitrogen. Some auxiliary fuel is generally burned to ensure complete combustion of the hazardous waste, for startup and for maintaining a minimum input of fuel. Some industrial waste has enough heat value to generate steam and thus achieve cost savings.

Successful compliance with emission limits has been achieved, and incineration should be considered for cleanup and reuse of contaminated material. Asphalt plants and cement kilns processing contaminated aggregates and soil have been dramatically proven to be in compliance with emission limits, with a scrubber or filter dust collector or both following the incinerator.

Filter collectors

These collectors use porous media to separate fine dust particles from the airstream at efficiencies up to 99.99 percent. Their technology will be covered in subsequent chapters more completely.[12,13] Figure 1.6

Figure 1.6 Cyclone and nine reverse-air filter collectors, tangential dusty air inlets, rotary air locks for bottom dust discharge, clean-air return ducts to plant for energy savings from all filter collectors, clean boiler smokestack, with steam from wood-fired boiler rising in background.

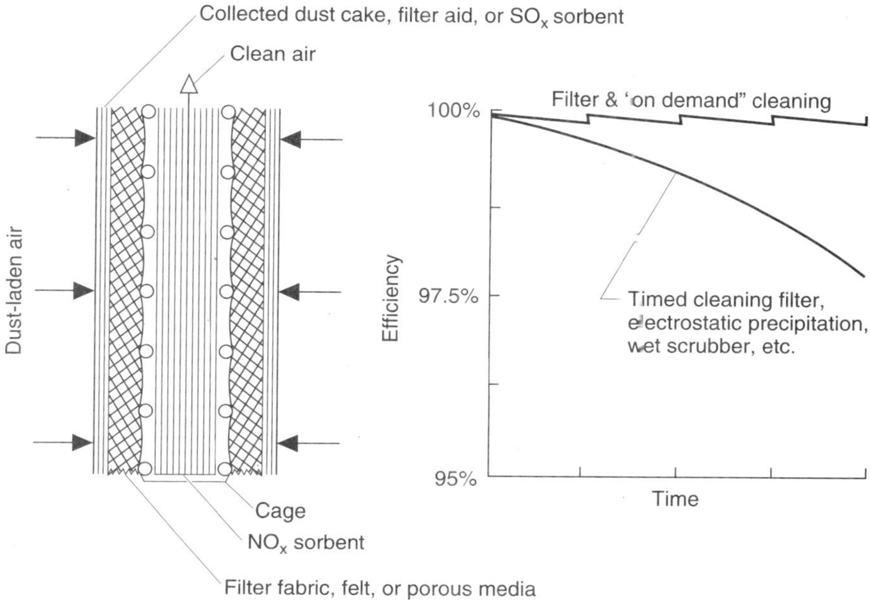

Figure 1.7 Filter separation with principle of operation in cross section, possible SO and/or NO sorbent locations, and chart comparing typical timed versus on-demand pulse-jet cleaning levels of collection efficiency.

illustrates a multiple filter collector installation, where savings are gained from the return of heated air to the plant in the winter or cooled air in the summer, from steam generated by burning the combustible wood waste in a boiler, and from collected product that may be recycled to improve plant yield.

Advancements have produced filter collectors that can control pollutants such as SO_2 and NO with sorbents in a "SO_x-NO_x-Rox-Box" (SNRB) combination (see Fig. 1.7). Where pressure or vacuum levels may exceed 1 psig, 27.7 in WG, a round housing design is generally used for its inherent greater strength to weight or hoop strength as seen in Figs. 1.8 to 1.11. A filter collector is often added after a primary collector to improve collection efficiency significantly, as shown in Fig. 1.12.

References

1. Suzanne A. Shelley, "Source Reduction: A Missed Opportunity," *Chemical Engineering—Environmental Engineering Supplement,* Nov. 1992, p. 3.
2. North Carolina Pollution Prevention Challenge Grant, March 17, 1989, State of North Carolina, Gov. James G. Martin.
3. *Industrial Ventilation Manual,* American Conference of Governmental Industrial Hygienists, 19th ed., 1986, pp. 11–34.

Figure 1.8 Round pulse-jet filter collector for 5-lb/in^2 gauge service, 24 filter bags 6-in diameter by 8-ft long, for abrasive powder service, two top doors, air header, diaphragm valves, and solenoid valve enclosure visible.

Figure 1.9 Round filter collector in fabrication shop, bottom cone hopper discharge opening, hopper port, and five-level sensor mounting couplings visible.

Figure 1.10 Round filter collector, one top door open with silicone gasket seals, wide spacing of filter bag holes, and two blow pipes visible over bag holes.

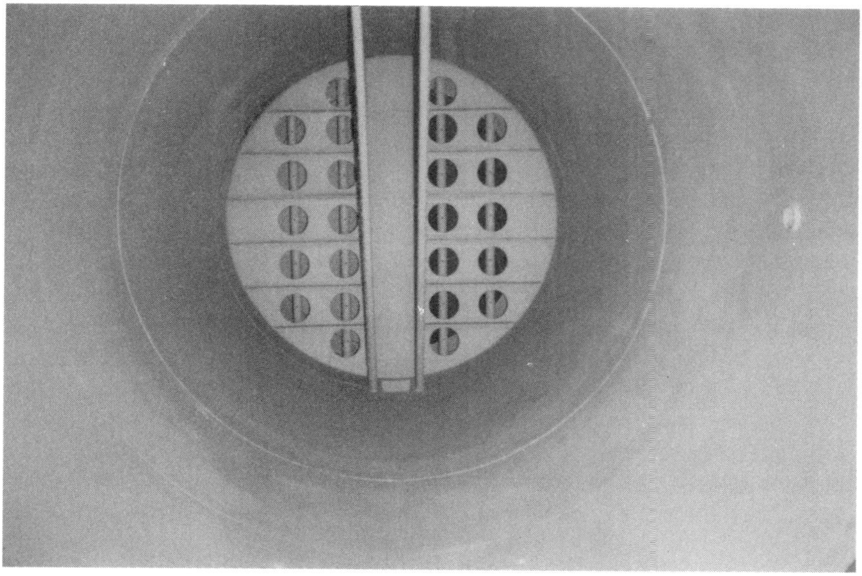

Figure 1.11 Round filter collector, 24 filter bag holes in tube sheet, blowpipes above tube sheet with pulse-jet holes centered over filter holes, 6-in round dirty-air inlet just below tube sheet with airstream directed between two vertical diffuser baffles of abrasion-resistant steel, with top-level sensor coupling visible in right housing wall.

Figure 1.12 Rectangular pulse-jet filter collector during installation, with prime paint coat, previously existing settling chamber, or dropout box, installed on left in dirty-air duct for primary dust collection prior to filter collector inlet, with round transition elbow to reduce hopper inlet velocity below 2000 ft/min.

4. U.S. Environmental Protection Agency, "Air Pollution Control Equipment for Particulate Matter," Research Triangle, NC, 1986, p. 116.
5. *Dust Collector Catalog,* DCE, Inc., Jeffersontown, KY.
6. Paul N. Cheremisinoff, "Fine Particulate Control in Air Pollution," p. 25.
7. Cheremisinoff, p. 41.
8. American Institute of Chemical Engineers, "Particle Size Classifiers," *Equipment Testing Procedure,* 2d ed., p. 31.
9. U.S. EPA, p. 135.
10. Cheremisinoff, p. 97.
11. "Economics of Fabric Filters vs. Precipitators," Electric Power Research Institute, Palo Alto, CA, June 1978.
12. Miles L. Croom, "Effective Selection of Filter Dust Collectors," *Chemical Engineering,* July 1993, pp. 86–91.
13. U.S. EPA, p. 106.

Chapter

2

Filter Collector Technology

"The positive collection mechanism of a filter baghouse ensures virtually complete collection of almost any dust or fume."

EPA
"Air Pollution Control Equipment for Particulate Matter."[1]

This book will aim to provide a concise and objective reference on filter collection technology, as well as a guide for their optimum design, sizing, specification, application, system accessories, performance, testing, and upkeep. It is intended to help in the design of optimum solutions for air pollution control, for estimating budget costs, for meeting regulations and emission limits, and for maintaining top performance. It is also intended to provide a ready reference when prompt compliance action is needed.

2.1 General Applications

This chapter explains that industrial dust collectors need to be as cost-effective as possible and now need to be even more highly efficient on small particle sizes, with the new CAAA and PM-10 fine particle regulations (EPA Clean Air Act Amendments and particle micron size collection under 10 μ). "When high collection efficiency on small particle size is required, however, the most widely used method consists of separating the dust from the air by means of a fabric filter."[2] They can also maintain a dependable level of operation over long periods of time.

Filters are being used to collect particles down to 0.3-μ size and smaller at efficiencies of 99.99 percent and better. Filter media are readily available for operation at temperatures up to 500°F (260°C). Higher temperatures previously required the use of heat exchangers or process changes to prevent exceeding the filter's safe maximum, or a different type of collector had to be used.

But new developments in ceramic and metal filter media promise to extend the safe operating temperatures to 1000°F (540°C) and beyond. Reduction or removal of air toxins as well as dust particles is being accomplished at these temperatures with absorbent additives for new multiple pollution control such as the SNRB (SO_x-NO_x-Rox-Box), currently in use and under development in power plants and operating at an optimum temperature of 850°F.[3]

Studies of comparative collection efficiency for various methods (see Fig. 1.2)[4] show the filter to be the most efficient in the fine-particle size range of 0.5 μ and up. At 1.0 μ and smaller the venturi scrubber efficiency drops sharply. The filter and electrostatic precipitator have essentially equal efficiencies down to 0.9 μ. Under that size the filter maintains the same efficiency while the precipitator drops off. The precipitator must be increased in physical size or footprint to match the filter collection efficiency for particle sizes below 0.5 μ.[5]

New developments

A compact hybrid particulate collector (COHPAC) met or exceeded emissions performance objectives for outlet opacity, fine-particle capture after an electrostatic precipitator, and SO_2 sorbent collection. In this new development filter collectors are being used as the final stage of particulate collection following an electrostatic precipitator on electric-generating power plant exhaust for maximum emission control.[7]

The use of pulse-jet filters as a performance upgrading secondary collector, following precipitators or a primary filter collector, shows a lot of promise due to new developments in their technology. Some advantages are

- Better efficiency on fines than any other industrial collection method, which should also prove valuable in future control of air toxins.
- Lower capital costs than other methods for equal efficiency.
- Lower operating costs than other methods.
- Less space required than other methods.
- Low-pressure, high-volume pulse-jet filter cleaning, at 8 to 12 lb/in^2 gauge, versus 60 to 100 lb/in^2 gauge normally.

- Longer filter life with the lower pressure cleaning, with service life up to 4 years.
- High filtration velocities, up to 15 ft/min, versus 2 to 5 ft/min on conventional filter collectors; since the dust load is greatly reduced by the primary collector, particles may carry some residual charge, and the low cleaning pressures reduce wear on filter membranes and media.
- Separate collection of excess sorbent for SO_2, with much of the dust collected in the upstream primary collector.
- Higher-temperature filter media and collector components for 800 to 1000°F (427° to 538°C) operation in current development.
- New filter configurations, such as oval, rigid porous, horizontal arrangement, cartridge, and others.
- Longer filter lengths up to 20 ft on 6-in diameter ceramic and 36 ft on 12-in needled felt, permitted by perforated cage-diffuser designs that provide more uniform cleaning effectiveness throughout the filter length.
- This cage-diffuser design also provides better filter media support for fragile high-temperature materials.
- This cage diffuser provides a supporting structure for NO_x reduction catalysts or sorbents with optimum exposure to the airstream after filtration cleaning of both the dust particulate and SO_x vapors.[3,6]

The world's largest pulse-jet fabric filters will be used on a new power plant to be completed in 1996 for a South African electric utility. This will be the first use of a primary pulse-jet filter on such a large, new coal-fired power plant. The filter for each of three 670-megawatt (MW) boilers will be designed for a gas flow of 2,754,000 actual ft^3/min (13,994 actual m^3/s), at 3.94 ft/min filtration velocity, guaranteed bag life of 20,000 h, and with emissions performance levels guaranteed. This decision was based on successful operation at two installations, one in Australia and one in South Africa, where a filter life of 4 years has been achieved using low-pressure, high-volume cleaning, with adequately high collection efficiency.

Similar size secondary filter collectors are now planned for a Texas utility system following electrostatic precipitator primary collectors. These will also be of the low-pressure pulse-jet cleaning design.

Secondary filters following a primary filter collector are being used more frequently on particle sizes down to 0.1 μ at efficiencies up to 99.999 percent. High-efficiency particulate arrestors (HEPAs) or absolute filters are used to separate particles in 0.1-μ sizes and finer at that high efficiency. They are used for meeting tighter emissions limits, for filtering air for clean rooms, for air makeup return for personnel breathing, for clean air supply for a process or combustion, and as a safety monitor or backup filtration after the primary filter. Secondary or even tertiary banks of filters thus extend the application range to finer-particle collection (see Chap. 5).

2.2 Filter Separation Mechanisms

The filter media can be a flexible fabric, needled felt, membrane, surface coating, or a composite of these; alternatively it can be a rigid porous cartridge or foam of paper, felt, plastic, ceramic, surface coating, or a sintered metal. All filter media use one or several mechanisms to actually separate submicron dust particles from the gas or airstream, where the open spaces in the filter media may be much larger.[8,9]

1. *Sieving.* Catching any particle size larger than the openings between the elements of the filter, like a sieve or screen.
2. *Interception.* Interrupting the path of particles when they pass within half their diameter of a filter element.
3. *Impaction.* Direct collision with a filter element.
4. *Diffusion.* For very small particles in the 0.2-μ size and below, random Brownian motion occurs and increases the chance of contact with the dust cake and filter elements and becomes the predominant means of filtration.
5. *Gravitational.* Settling of particles on the filter media or dropout into the hopper below.
6. *Electrostatic.* Attraction that may arise, or may be set up intentionally, between the dust and the filter to pull particles into the filter or dust cake, cause particle agglomeration, and improve filter cleanability.
7. *Filter dust cake.* Process dust particles that build up on the filter surface and provide a thicker mat or bed and finer flow path than the media itself. This retention of filter cake provides for collection of smaller particles, reduction of fines penetration into the filter, and usually lowers pressure drop across the filter. Therefore, filter cleaning may need to be delayed on initial startup or should be less aggressive and less frequent to permit the retention of an optimum layer of dust cake at all times.
8. *Precoat or flow aid cake.* An intentional layer of process dust or loosely porous flow aid powder that is built up on the filter surface

initially and/or during operation like the dust cake. This powder mat or layer will filter finer particles, reduce penetration of fines into the filter media, reduce pressure drop across or through the filter, and the powder selected can be a neutralizing agent to reduce SO_2 or other contaminants.

2.3 Flow Aid or Filter Dust Cake

Precoating or continuous coating with a dust cake of the particulate to be collected, or with a flow aid or neutralizing powder, can improve filtration efficiency especially on the finest particle sizes, as discussed in parts 7 and 8 in the above list. Precoating with process dust will usually be begun at a reduced feed rate prior to operating at the full dust input rate (see part 8).[8]

Flow aid powder feed will usually be done with an auxiliary injection system that feeds a uniform amount of a powder such as lime accurately into the airstream inlet of a filter collector continuously during operation (see part 8 of list). The powder will need to be dispersed evenly throughout the airstream to produce a uniform coating over the filter elements. The feed rate may need to be controlled in proportion to the inlet dust loading, airstream flow, pressure drop across the filters, or level of SO_2 or target toxic pollutant in the airstream or exhaust. This will be discussed in detail in Chap. 7.

Filter dust cake is commonly built up during the first 2 to 6 h of operation. The main system fan can be used to deliver a reduced dust loading of 10 to 20 percent of the normal feed rate. If the process dust has a fairly broad range of particle sizes, it should make a good filter cake. If it is uniform in size and is fine, then a separate filter aid powder should be considered that has a broad size range. Resistance to flow can be reduced by several inches water gauge and filtration efficiency can be improved by providing this more porous filter cake bed.[9]

Reverse-air cleaning filter collectors serving large airstream volumes in electric utility powerhouses have probably used dust cakes more than any other application. The benefits that have prompted extensive evaluation studies and operational development include

- Less penetration of fines into the filter media
- Reduced pressure drop across the filters
- Resultant lowering of fan energy costs
- Longer service life of the filters, which is critical in maintaining continuous generation of electricity
- Cost savings from less frequent filter replacement

- Neutralization of SO_2 emissions in combustion exhaust through the use of lime or other sorbent in the filter aiding powder
- Reduced corrosion of downstream equipment

Pulse-jet, shaker, and reverse-air collectors should all benefit from process dust and flow aid powder coatings in most applications. Initial precoating with process dust would be worth consideration on every filter collector as a standard start-up procedure (see Chap. 11), and flow aid powder coating should also be evaluated for its possible benefits, especially where

1. Fine particles must be collected
2. Filter cleaning is by any powered method
3. Airstream flow exceeds 1000 ft^3/min (28 m^3/min)
4. Energy costs can be saved by reducing fan power
5. Filters are relatively expensive, easily damaged, or blinded by penetration of fines
6. Neutralization of toxic pollutants is also necessary

2.4 Common Deficiencies

Selection, design, and application of collection systems will often suffer from two common deficiencies:

1. The most frequent is a failure to take into account the normal variation in dusty airstream characteristics and volume during varying operating conditions. Careful design will consider maximum, minimum, and expected normal rates. The Environmental Protection Agency (EPA) comments, "On the average, nothing is average."[10]

2. Another is being too optimistic about equipment performance, characteristics of the dust and airstream, and cost estimates. It is better to be a little conservative in preliminary sizing and cost estimating, especially on new applications where the temptation is not to kill the project by using realistic costs; then actual performance may be better and costs lower than promised.

In pollution control the project *must* meet the regulated emission limits. So budgeted costs have to adequately cover the necessary equipment for operation as promised. Where savings are expected, accurate cost and performance figures need to be developed to measure and justify the proposal. For subsequent follow-up and verification, actual results should be on the mark or beat those promised.

2.5 Sizing

Process volumes

Determine the minimum and maximum airstream volumes and dust loading capacity needs by volume or weight. Often this will already be determined by the project operating levels. Another frequent basis for sizing is the estimated air volume needed to capture the expected dust loading at all the plant emission sources. This volume can be calculated from accumulating the volume of individual capture hoods and ductwork inputs to the system required as described in Chap. 8.

Filtration velocity V_f or air-to-cloth ratio A/C

This is critical for sizing a filter collector and for its ultimate performance. The basic data should be derived from tests for new dust products, or past experience on similar dust and filter media applications, and then a preliminary filtration velocity can be selected. This is often stated as the air-to-cloth ratio in ft^3/min per ft^2, which reduces to ft/min of filtration velocity. Typical operating filtration velocities in industrial applications may range from 2 to 8 ft/min with a maximum up to 15 ft/min. The pressure drop or resistance of the clean filter material is measured at 0.5 in WG (1.269 to 3.807 g/cm^2) for the ASTM permeability standard of air volume in ft^3/min per ft^2, and will typically run from 10 to 50 ft^3/min per ft^2 of filter media.

Sizing the filter area. The gas or airstream volume Q in cubic feet per minute divided by the filtration velocity V_f in feet per minute determines the required filter area in square feet:

$$\text{Filter area} = \frac{Q}{V_f} = \frac{\text{air volume}}{\text{filtration velocity}}$$

$$= \frac{10{,}000\ \text{ft}^3/\text{min}}{5\ \text{ft/min}}$$

$$= 2000\ \text{ft}^2 \qquad (2.1)$$

$$\text{Metric area} = \frac{283.2\ \text{m}^3/\text{min}}{1.5234\ \text{m/min}}$$

$$= 185.9\ \text{m}^2$$

$$\text{(Rechecking)} = (185.9\ \text{m}^2)(10.76\ \text{ft}^2/\text{m}^2)$$

$$= 2000\ \text{ft}^2$$

Table 2.1 lists nominal filtration velocities for selected dust materials for both low-pressure shaker or reverse-air cleaning and for high-pressure pulse-jet compressed air cleaning. These will usually run from 2 to 15 ft/min per ft^2. Figures for a single value or a range of bulk density are listed for each material also. Before this selection and a collector size calculation is made, some other factors should enter into the evaluation; therefore, more detailed evaluation is presented in Eq. (2.2). Effective filtration velocity equals nominal filtration velocity times factors (for application, temperature, particle size, and dust load):

$$\begin{aligned} V_{fe} &= V_{fn} \times A \times T \times P \times D \\ &= 8.0 \text{ ft/min (cement)} \times 0.9 \times 0.8 \times 1.0 \times 0.8 \\ &= 4.6 \text{ ft/min} \qquad \text{(for cement as an example)} \end{aligned} \tag{2.2}$$

where V_{fe} = effective filtration velocity and V_{fn} = nominal filtration velocity from Table 2.1 and the factors are as follows:

A = application	0.8	Oily, moist or agglomerating
	0.9	Product collection
	1.0	Nuisance dust collection
T = temperature	1.0	Up to 110°F (43°C)
	0.9	From 110 to 225°F (107°C)
	0.8	Above 225°F (107°C)
P = particle size	0.8	Under 3.0 μ
	0.9	From 3 to 9 μ
	1.0	For 10 to 50 μ
	1.1	For 51 to 100 μ
	1.2	Above 100 μ
D = dust load	1.2	For 10 or less gr/ft^3
	1.1	For 10 to 20 gr/ft^3
	1.0	For 20 to 50 gr/ft^3
	0.9	For 50 to 80 gr/ft^3
	0.8	Above 80 gr/ft^3

Can or filter passage velocity V_c

The airstream velocity on entering the passage areas between the filters, often called *can velocity,* might better be called the filter passage velocity. This describes more exactly the flow function involved, which is the airstream velocity passing around the filters. A general top average is 200 ft/min, but since "on the average, nothing is average," a more detailed explanation is needed.

Can velocity is usually an upward vertical flow in filter collectors that are cleaned on-line, so it must be limited to a velocity that will not reentrain the dust particles that have been cleaned. These

TABLE 2.1 Filter Sizing Factors for Primary Filter Collectors

Material (dry dust)	Filtration velocity V_f by type of cleaning: Low pressure, ft/min	High pressure, ft/min	Maximum can velocity V_c, ft/min	Bulk density at rest, lb/ft^3
Abrasive dust	3.0	10.0	225	120–150
Activated carbon	2.0	6.0	175	20
Adipic acid	2.0	7.0	200	40–50
Alumina	2.5	8.0	200	40–55
Aluminum dust	3.0	9.0	200	40–65
Aluminum oxide	2.0	9.0	200	60–90
Ammonium phosphate	2.0	7.0	200	55–80
Asbestos	3.0	8.0	175	20–40
Baking powder	2.5	9.0	175	40
Bauxite	2.5	8.0	200	45–75
Bronze powder	2.0	8.0	200	70–90
Calcium	2.0	6.0	175	30
Carbon, black	2.0	6.0	150	10–35
Carbon, Banbury mixer	2.0	7.0	175	20–40
Cement	2.5	8.0	225	65–95
Ceramic pigments	2.5	8.0	200	50–90
Charcoal	2.5	5.0	175	18–25
Chrome ore	2.5	9.0	225	125–140
Clay dust	2.5	9.0	200	50–70
Clay, kaolin	2.5	9.0	200	45–55
Coal, pulverized	2.5	8.0	175	35–50
Cocoa, chocolate	2.8	10.0	175	30–35
Coke	2.5	7.0	175	30–40
Corn flour	3.0	8.0	200	45–55
Cotton	3.5	8.0	175	15–30
Cosmetics	1.5	10.0	200	45–90
Dacron flakes	2.0	6.0	175	30–40
Diatomaceous earth	2.2	10.0	150	5–18
Dyes	2.0	7.0	200	25–40
Enamel frit	2.5	9.0	200	80–100
Epoxy powder	2.0	8.0	200	45–55
Feeds, grain	3.5	14.0	200	30–40
Feldspar	2.2	9.0	225	60–75
Fertilizer	3.0	8.0	200	35–55
Fiberglass, chopped strand	2.5	8.0	175	15–25
Flour	3.0	12.0	200	35–50
Fluorspar	3.0	8.0	225	85–110
Fly ash	2.5	7.0	200	35–65
Fly ash, COHPAC secondary	4.0	15.5	200	35–65
Glass	2.5	5.0	200	65–105
Grain	3.0	10.0	200	35–40
Granite	2.5	8.0	225	80–100
Graphite	2.0	5.0	200	28–55
Gypsum	2.0	9.0	200	50–80
Iron ore	3.0	11.0	250	120–180
Iron oxide	2.5	7.5	200	75–95

TABLE 2.1 Filter Sizing Factors for Primary Filter Collectors (*Continued*)

Material (dry dust)	Filtration velocity V_f by type of cleaning: Low pressure, ft/min	High pressure, ft/min	Maximum can velocity V_c, ft/min	Bulk density at rest, lb/ft^3
Iron sulfate	2.0	6.0	200	60–85
Lampblack	2.0	6.0	150	35–45
Lead oxide	2.0	7.0	225	50–70
Leather dust	3.5	10.0	175	15
Lime	2.5	9.0	200	35–60
Limestone	2.7	8.0	200	60–70
Manganese	2.3	7.0	200	70–80
Marble	3.0	8.0	225	80–95
Metallurgical fumes	1.5	6.0	150	10–20
Mica	2.7	9.0	175	13–30
Milk powder	4.0	8.0	200	15–35
Nylon flakes, powder	2.0	6.0	175	30–40
Oyster shell	3.0	8.0	200	50–80
Perlite	3.0	9.0	175	10–20
Phosphate rock, ground	3.0	9.0	225	60–90
Plastic pellets	3.5	10.0	225	30–45
Plastic powder and resin	2.5	7.0	200	25–45
Potassium carbonate	2.5	7.0	200	50–80
Quartz	2.8	9.0	200	85
Refractory dust	3.0	9.0	200	60–85
Rock dust	3.0	9.0	225	80–90
Rubber dust	3.0	9.0	200	25–50
Salt	3.0	9.0	200	50–85
Sand	2.5	10.0	225	90–120
Sandblast dust	2.5	9.0	200	80–90
Sawdust, dry wood	10.0	17.5	175	10–25
Sawdust, fines	4.0	10.0	150	10–20
Silica	2.5	8.0	225	80–95
Slate	3.5	10.0	225	80–100
Soap, detergent	2.0	8.0	175	15–35
Soda ash	3.0	9.0	200	50–65
Spices	2.7	10.0	200	30–50
Starch	3.0	7.0	200	25–50
Steel, machine dust	3.0	10.0	225	80–150
Sugar	2.0	8.0	200	40–60
Talc	2.5	9.0	200	35–45
Titanium dioxide	2.0	7.0	175	45–50
Tobacco, ground	3.5	13.0	200	10–30
Urea formaldehyde	2.5	8.0	175	35–45
Wheat flour	2.0	10.0	175	30–45
Wood chips and dust	7.5	17.5	200	15–35
Zinc, metallic dust	2.5	8.0	250	200
Zinc oxide	2.0	7.0	200	25–55
Zirconium oxide	2.5	7.0	225	70–85
Intake filter-compressed air	4.0	—	—	—

cleaned particles need to drop out by gravity. In other words, the collected dust must be able to fall out through this upward flow during filter cleaning.

If the fine dust cannot drop out, it will be driven back up onto the filter surface or reentrained. This is a problem on tall filters where the fines are retained on the upper part of the filters indefinitely. The area that is effectively cleaned and thus available for filtering is reduced, which makes the filtration velocity increase, which in turn increases the pressure drop across the filter.

Filter collectors that are set up to clean one module or a portion of the total collector off-line at one time should not experience the reentrainment problem. In this case the upward flow is stopped during cleaning of each module, so the dust is free to fall. Then a sequential timer control will advance to the next module according to the cycle time setting and thus all modules are cleaned in sequence but just one at a time. This is common practice on large reverse-air or shaker collectors and on some pulse-jet collectors.

Upward passage or can velocity is listed for each of a number of dust materials in Table 2.1. These are only guides and can vary if the dust characteristics change. Again, past experience or tests should help in the adjustment of the listed nominal velocities.

Pressure drop

The pressure drop or resistance in a filter collector is primarily related to the level of the following factors:

- Filtration velocity V_f, or A/C ratio
- Filter media selected
- Dust loading in the airstream
- Dust fineness and characteristics
- Can or passage velocity V_c
- Filter cleaning effectiveness, or how close the cleaning process comes to "new" filter condition
- Amount and porosity of the dust cake retained on the filters
- Mechanical draft or pressure drop due to flow through the collector enclosure and the inlet and outlet transition designs
- Filter drag characteristic S, in inches water gauge per feet per minute or resistance or pressure drop of the particular media relative to the filtration velocity

There are theoretical formulas for calculating pressure drop across a filter collector. However, they generally involve an experimental factor for the filter drag S, in inches water gauge per feet per minute, other coefficients, or a resistance factor. Unless these factors can be accurately determined, this puts the calculation right back on an empirical basis. The factors must also reflect the variable operating conditions.

So the expected pressure drop is usually estimated either from the typical pressure drop for a similar application or from an estimated sum of pressure drops across the filters usually referred to as the *filter drag.* To this must be added the mechanical draft losses of the filter housing, hoods, and ductwork.

There is a direct relationship between pressure drop P_c (ΔP) through a filter collector, and drag resistance S across the filter media times the filtration velocity V_f through the filter media. This relationship is expressed as pressure drop = filter drag × filtration velocity.

$$P_c = S \times V_f \tag{2.3}$$

Relatively speaking

- Filter drag S in inches water gauge per feet per minute, is the resistance to flow.
- Filtration velocity V_f (air-to-cloth ratio), in feet per minute, is the airstream velocity through the filter media at design or operating conditions.
- Pressure drop P_c (ΔP), in inches water gauge, is the total operating pressure drop or resistance across the filter collector and this ΔP increases as the product of the filtration velocity times the drag or resistance increases.

For example, for a typical industrial collector, the pressure drop would be calculated as shown in Eq. (2.3).

This system pressure drop can usually be a fairly realistic figure. As an example, in a COHPAC pulse-jet filter as mentioned above, the drag was only 0.4 in WG/(ft/min) times a V_f (filtration velocity) of 15 ft/min = 6 in WG due to a very light dust loading. In a typical industrial process collector, the mechanical drag losses were measured at 2.5 in WG/(ft/min) times the filtration velocity V_f of 4.0 ft/min so the total pressure drop was 10.0 in WG. (*Note:* For fan sizing, the ductwork losses would have to be added, so if they were 5 in WG, the total fan output volume would need to be available at 15.0 in WG.)

$$\begin{aligned} P_c &= S \times V_f \\ &= 2.5 \text{ in WG/(ft/min)} \times 4.0 \text{ ft/min} \\ &= 10.0 \text{ in WG} \end{aligned} \tag{2.3}$$

In one COHPAC secondary collector development case reported already, filter drag S = 0.4 in WG/(ft/min) and filtration velocity V_f = 15 ft/min, so

$$P_c = S \times V_f = 0.4 \text{ in WG/(ft/min)} \times 15 \text{ ft/min}$$
$$= 6.0 \text{ in WG} \qquad (2.3)$$

Filter drag S measurements should be available from suppliers of filter materials for expected dust cake conditions in operation, or a reasonable estimate developed from experience. However, tests may be necessary to determine dependable drag levels for sizing at the expected operating conditions, or the bid specifications can be written on the best experience or estimates. A performance guarantee can be written into the specifications to be based on operating results as determined by test measurements.[11] Some adjustment in total purchase price could be agreed upon based on operating test results.

As one source recommended, "Be careful of promises to provide low pressure differentials with high filtration velocities (air-to-cloth ratios). This is contrary to the laws of physics. Such promises may lead to undersizing the fan and the collector, thus limiting the capacity in operation."[4]

Filter media

The media is the key to filter collector performance. All other design and application factors hinge upon the filter media that actually separates the dust from the airstream. The selection of filter material and configuration should be based upon experience, testing, new development evaluation, and performance guarantees to determine the best material for the specified operating conditions (Table 2.2). Search for the optimum combination of performance and efficiency at the lowest cost!

- Permeability is one of the primary design and operating parameters of filter media. This is usually measured in cubic feet per minute of airflow per square foot of new or clean filter at a pressure differential of 0.50 in WG across the filter, by ASTM standard D-737. The usual permeability range is from 10 to 40 ft^3/min per ft^2 at the pressure drop of 0.50 in WG, or the equivalent feet per minute. *Permeability* is related to filtration velocity (air-to-cloth ratio) and filter drag.
- *Filtration velocity* is the measure in feet per minute of actual flow, or maximum actual design flow, through the filter in operation with expected dust cake present at the operating conditions and pressure drop [equal to V_f in Eq. (2.3)]. *Air-to-cloth* ratio is another

TABLE 2.2 Filter Media Selection

Material	Temperature limit, °F	Chemistry acid/base	Residual moisture	Service strength		Filtration efficiency	Permeability, $(ft^3/min)/ft^2$	Drag, in WG/(ft/min)	Cost, $\$/ft^2$
				Flexibility	Cleaning				
Felts									
Polypropylene	190	E/E	P	E	G	E	30+	0.5	1.00
Polyester	275	G/F	F	F	E	E	30+	0.5	1.00
Acrylic	260	V/F	E	G	F	E	30+	0.5	1.00
Nomex	400	G/G	G	G	G	E	30+	0.5	3.00
Ryton	375	E/E	E	E	E	E	30+	0.5	3.20
P 84	460	E/P	E	E	E	E	30+	0.5	3.30
Teflon	475	E/E	E	F	P	E	30+	0.5	5.20
Woven									
Fiberglass	500	G/P	E	F	P	G	20+	0.7	2.50
Ceramic	1200	E/G	G	G	P	E	30+	0.5	7.00
Metals	1000	F/G	E	G	E	G	20+	0.7	6.50
Membrane (expand porous surface filtration coating)									
PTFE	500	E/E	E	F	E	E	30+	0.6	2.50
Acrylic	275	V/F	E	G	G	E	30+	0.5	1.00
Finishes (thermally singed, or chemical finish)									
Graphite grounding	500	E/E	E	G	G	E	30+	0.5	1.50
Silicone	500	E/E	E	G	G	E	30+	0.5	1.00
Teflon	500	E/E	E	G	G	E	30+	0.5	1.50
Ground wires	500	G/G	G	G	G	E	NA	NA	0.50
Epitropic grounding	400	G/G	G	G	G	E	—	—	0.50
Rigid cartridge									
Cellulose/ synthetic	240	F/F	F	G	E	G	15+	0.9	2.00
Polyester	275	G/G	G	G	G	E	30	0.5	2.50
Nomex	400	G/G	G	G	G	E	30	0.5	5.50
PTFE membrane	400	E/E	E	G	G	E	25	0.7	2.50

Note: E = excellent, V = very good, G = good, F = fair, P = poor.

way of expressing the filtration velocity as the volume of airflow in cubic feet per minute per square foot (feet per minute) of filter area in operation with dust cake present at the operating pressure drop and conditions [equal to V_f in Eq. (2.3)].

- *Filter drag S* is the measure in inches water gauge of the pressure drop across a filter at the filtration velocity in feet per minute at the operating conditions with dust cake present [S in Eq. (2.3)]. With new or clean filter media and at the ASTM standard conditions, the drag should be proportional to the permeability measurement at air flow volume per ft^2 at 0.5 in WG.

Probably the most significant improvement in filter performance in recent years has been the development of needled felt materials. Woven fabric is still used on some low-pressure cleaning applications or for high temperatures using fiberglass, ceramic, or metal fabrics. Felts have improved most industrial applications at less than 500°F (260°C) through higher filtration efficiency on much finer particle sizes of dust. Efficiency has been increased from around 98 up to 99.99 percent levels. While pressure drop across clean filters has been maintained about the same in many cases, the operating drag has generally been reduced for a big benefit. This has been due to reduced penetration of fine particles into the felt media compared to woven fabric filter media.

Felted media bag life has been lengthened by 50 to 100 percent over most fabric types due to the reduced penetration which results in less abrasion and wear of the filter fibers. Finer particles down to 0.3 μ can be separated before penetration and blinding of the filter becomes severe. Felt is more suitable for a finish treatment, coating, or especially the application of a membrane for the most effective filtration of fines. Energy costs for the fan drive have been reduced, due to the generally lower drag, or pressure drop, in operation.

Currently, more exotic materials are being adapted to extend the temperature limits, resistance to chemical attack, and performance of filters. When the maximum safe temperature can be raised in the collector, the overall system cost, both initially and for energy consumption, can be reduced by elimination of heat exchangers, air dilution, or other process cost changes. Ceramic and metal fabrics, and rigid foam and sintered porous materials, are being developed and have been successfully operated at up to 1000°F (538°C).

Cartridges

This special filter construction is made by pleating or folding the filter media in multiple accordion folds which permits a much greater filter area to be used in the same space as a filter bag. One of the advantages of cartridge collector design is that usually less headroom and floor space is required than for bag filters. The earlier cartridges used as many pleats as possible, but the close spacing often caused trouble through the buildup of fine or sticky materials, with rapid clogging of the pleats and poor cleanability. Designs from 50 to 320 folds around the circumference are the usual range used to capture the dust. Washable cartridge filters, open pleat construction, and electrostatically conductive filter cartridges for grounding are among the new developments that are available now.

Case Study On a fine lime dust collection application using 320 pleats per cartridge, the lime dust would build up in the pleats and combine with moisture in the air to form hydrated calcium hydroxide. This made a hard cake that would not release easily from the cartridge, with a loss of effective filter area and throughput capacity, so frequent off-line cleaning or replacement was required. New polyester cartridges with a wider pleat spacing by using only 50 pleats per cartridge, and with a Teflon membrane outer surface were the replacement selection. The lime was then easily released and cleaned off, the filtration velocity has been raised, and the collector operates at a lower pressure drop.

A recent cartridge development is a patented pleated filter that provides up to 45 ft^2 of filter area in a compact 36-in-long cartridge shape. The media is washable and achieves collection efficiencies of 99.9 percent, with pressure drops equal to or less than those of equivalent bags. Several developments in higher-temperature media have extended cartridge use to 500°F (260°C) dependably and higher temperatures may be met by current developments.

References

1. U.S. EPA, "Air Pollution Control Equipment for Particulate Matter," p. 106.
2. U.S. EPA, "Air Pollution Control Equipment for Particulate Matter," p. 106.
3. 3M, Minnesota Mining and Manufacturing Company, "Ceramic Filters—Nextel Filter Bags," 1991.
4. DuPont Company, Teflon B Filters.
5. U.S. EPA, p. 110.
6. "Economics of Fabric Filters vs. Precipitates," Electric Power Research Institute, Palo Alto, CA, June 1978.
7. Newsfront section, *Chemical Engineering,* August 1993.
8. U.S. EPA, pp. 106, 108–110.
9. U.S. EPA, pp. 111, 112.
10. U.S. EPA, p. 116.
11. U.S. EPA, pp. 112, 113.
12. L. S. Marks and T. Baumeister, *Standard Handbook for Mechanical Engineers,* McGraw-Hill, New York, 1967, pp. 7-80 to 7-90, 12-141, 12-142.
13. *Industrial Ventilation Manual,* American Conference of Governmental Industrial Hygienists, 19th ed., 1986, pp. 11-10 to 11-13.

Chapter

3

Filter System Design

"Don't put a low speed airfoil on a jet airplane."

Performance of a filter collector system is determined by the suitability of the design and specifications for the operating conditions, by the filter medium selection, and by the accessory equipment for the system. Chapter 2 covered filter technology and media selection, while later chapters will cover accessory functions and the importance of optimizing collector design. Chapter 8 provides guidelines for dust capture and conveying system design. Chapter 10 will cover costs and specifications. This chapter will develop guidelines for designing a suitable filter system and present case studies on a wide range of industrial applications for review.

System design is comparable to choosing an airfoil section for an airplane wing, selecting a propulsion system, and designing the aircraft configuration so a new plane meets or exceeds the performance desired. With the basic design specified, the performance levels are pretty well defined for lift, drag, load capacity, power required, costs, and speeds attained. Subsequent fine tuning can optimize these results, sometimes significantly.

The levels of collection system efficiency, pressure drop, energy use, physical size, and costs will be fairly well determined after the filter collector design has been decided on, the filter material has been chosen, accessory functions have been defined, and the performance and construction specifications have been written. Thus the need for early consideration of alternative collection technology and design in a project evaluation is apparent. A low-performance bin vent filter with on-off cleaning should not be chosen for a large collector system needing high efficiency and continuous performance, or vice versa.

During and after the collector design stage and evaluation of bids, the opportunity is available to optimize features. The decisions about many of the accessories required for the operation, such as the dust capture system or the best fan to move the air, may appear to be routine. But if the accessories are chosen carefully, big benefits in investment and operating cost savings can be achieved. So foresight in planning the total system can be valuable, rather than letting it be only a necessary afterthought!

One of the main considerations in filter system design is the choice of a cleaning method and inclusion of recent improvements. Many new design features, filter media, and alternative applications have been developed. So be on the lookout for advancements in collector system performance, filter temperature and effectiveness levels, particle size efficiency, service life and warranties, and cost comparisons.

3.1 New Developments

Some of the recent developments highlighted below are discussed in the case studies later in this chapter.

- Application of optimum design analysis to a filter collector so that it meets emission regulations and capacity requirements better and at lower cost than previous typical collectors.
- Off-line cleaning by collector modules when handling fine-particulate dusts.
- Higher-temperature bag and cartridge filter media for 1000°F (538°C) service and beyond.
- Perforated cage-diffuser to permit filter bag lengths up to 20 ft on 6-in diameter and 40 ft on 12-in diameter and location of sorbent inside this cage.
- Secondary pulse-jet bag houses using a filtration velocity up to 15.5 ft/min on fines collection after a precipitator or other primary collector.
- The combination of SO_2, NO_x, and dust collection in one enclosure operating at an optimum 850°F (454°C).
- Medium pressure, 8 to 40 lb/in^2 gauge, pulse-jet cleaning for longer bag life due to pressure lower than the 60 to 100 lb/in^2 gauge typical pressure.
- Electrostatic augmentation of filter collection.[2]
- World's largest pulse jet at over 2,750,000 actual ft^3/min airstream volume being installed for power plant fly ash collection.

- Availability of firms who take sole source responsibility for overall filter system design, emissions, efficiency, installation, and performance reliability.
- In July, 1993 the first reduction in acid rain pollutant levels across the United States was reported after 11 years of monitoring from 1980 to 1991 by the U.S. Geological Survey. Also in July 1993, the EPA reported a 13 percent decline in chemicals released into the air, at 2.01 billion lb, for 1991.[3]

3.2 Collection System Effectiveness

Particulate collection efficiency or effectiveness is usually measured either as a dust capture efficiency or as an emissions dust load, usually specified at definite particle size ranges by the applicable regulations. Opacity levels were formerly the primary concern and measurement, and many regulations still call for measurement and recording of opacity. But particulate, and also toxic, emissions should be better controlled by either direct monitoring of the specific particle or toxic emissions, or by periodic sample analysis of them for correlation with operating parameters. Then analysis can be made and action taken to reduce or remove the specific problem pollutants (see Chap. 7, Accessories). The two dust particle measures of collection are directly related as follows.

Emissions level

As an example, the inlet dust loading might be 50 gr/ft^3, the allowable emissions rate might be 0.02 gr/ft^3, and the actual operating emissions measured at 0.01 gr/ft^3. Then for a 10,000 ft^3/min collector the weight input rate is (10,000 ft^3/min $\times$ 50 gr/ft^3)/(7000 gr/lb) = 71.43 lb/min. The allowable emissions weight output would be (10,000 ft^3/min $\times$ 0.02 gr/ft^3)/(7000 gr/lb) = 0.0286 lb/min. And in this example the actual emissions weight output would have been (10,000 ft^3/min $\times$ 0.01 gr/ft^3)/(7000 gr/lb) = 0.0143 lb/min.

Dust collection efficiency

This is a measure of the relative ability to separate or collect the inlet dust load. The collected dust divided by the inlet dust load times 100 is the percentage efficiency in collection of the inlet dust loading. This can be calculated for the dust loading basis as follows:

$$\text{Collection efficiency} = \frac{\text{inlet} - \text{emission dust load}}{\text{inlet dust load}} \times 100\%$$

$$= \frac{\text{collected dust load}}{\text{inlet dust load}} \times 100\% \tag{3.1}$$

The allowable collection efficiency at 0.02 will be

$$\text{Allowable collection efficiency} = \frac{50 - 0.02}{50} \times 100\%$$

$$= \frac{49.98}{50} \times 100\%$$

$$= 0.9996 \times 100\%$$

$$= 99.96\%$$

Then the actual collection efficiency at 0.01 will be

$$\text{Actual collection efficiency} = \frac{50 - 0.01}{50} \times 100\%$$

$$= \frac{49.99}{50} \times 100\%$$

$$= 0.9998 \times 100\%$$

$$= 99.98\%$$

If the regulations allow a maximum emissions level of 0.02 gr/ft^3 and 99.96 percent efficiency, then the actual level of 0.01 gr/ft^3 and 99.98 percent efficiency is obviously better than required. If the regulations or operating guidelines needed the efficiency calculated on a weight basis, it is

$$\text{Allowable weight efficiency} = \frac{(71.43 - 0.0286 \text{ lb/min})}{71.43} \times 100\%$$

$$= \frac{71.4014}{71.43} \times 100\%$$

$$= 99.96\%$$

$$\text{Actual weight efficiency} = \frac{71.43 - 0.0143 \text{ lb/min}}{71.43} \times 100\%$$

$$= 0.9998 \times 100\%$$

$$= 99.98\%$$

In this example both emissions loading level and the collection efficiency are better than the regulations. This does not always happen easily, and then further control action or alternative design will be necessary, as will be shown in some case studies in this chapter. Emissions monitors will be discussed in Chap. 7.

3.3 Filter Cleaning Methods

On-off

On-off or collapse operation of a collection system fan or pneumatic conveyor blower will clean filters by the "sag of the bag" when the airflow is turned off. This is useful for intermittent operations and less critical collection. Bag, drum, mixer, truck or rail loading and unloading, and bin vent or conveyor transfer generated nuisance dust collectors can be so cleaned. If the operation and dust load permit, this is the lowest-cost cleaning system. Fabric bag filters are usually chosen. The on-off, or collapse, cleaning effect is of some benefit on any collector but obviously has little effect on continuously operating systems.

What must be considered is the on-off cleaning that takes place when the collector is turned off at the end of an operating period or when there is a power outage. The controls should provide for a timer delay of 5 to 10 min after the fan and collection equipment stop before shutting off the discharge conveyor or valve.

This should allow any dust material that drops off the filters after they collapse to be collected in the hopper and emptied out, assuming that there is no reason the material needs to be held longer in the hopper.

Shaker

Shaker cleaning maintains effective filtration by periodically shaking the filters manually, as cycled by timers, by photohelic pressure drop controls, or by operating control interlocks. For instance, a vehicle unloading system blower motor could be interlocked to run the shaker cleaning motor continuously whenever the blower is operating. Generally a motor-driven eccentric mechanism shakes all the bags simultaneously on small units, or a section at a time on larger collectors. On-off cleaning supplements shaker cleaning whenever the fan or blower is turned on or off and the filter bags collapse. Fabric filters are the primary media choice.

The small shaker-cleaned collector on a silo shown in Fig. 3.1 could be cleaned by on-off or shaker cleaning depending on the length of the loading cycle. A bucket elevator, shown on the left of the silo, will

Figure 3.1 Small shaker-cleaned filter collector on a storage silo fed by a bucket elevator.

usually run for an hour or more, so either an interlocked shaker cleaning or timed cleaning while the bucket elevator is running would be the better choices.

Recent use of acoustic horns to improve the flow of bulk solids out of collector hoppers has brought to light their benefit in acoustic shaker cleaning of filters. Some designs now use an acoustic horn as the primary shaking mechanism. Horns have also been added to improve filter cleaning on collectors of other designs that had cleaning problems (see Chap. 7, Accessories).

Large collectors may use shakers that only clean one or a few of the sections of the total filter area at a time. This maintains sufficient filter area on-line to hold down the filtration velocity and the pressure drop. There has been a trend away from shaker cleaning to reverse-air and pulse-jet cleaned collectors, due to their advantages of providing for higher filtration velocities, smaller physical size for the same airstream volume, lower wear and longer life of the filters, and higher collection efficiency.

Reverse air

Reverse-air cleaning uses a relatively low pressure, usually 2 to 4 lb/in^2 gauge, and high-volume air supply to clean the filters in sequence. Air pressure of 6 to 15 lb/in^2 gauge can be supplied by a centrifugal or positive displacement blower for cleaning. Centrifugal fans have been used for reverse-air cleaning at the low pressures of 2 to 4 lb/in^2 gauge.

Woven fabric filters have been widely used, but needled felt media is replacing woven fabrics to a large extent. Felt helps reduce system pressure drop, energy cost, and particle penetration; is more suitable for the use of a membrane on the filter surface; and improves collection efficiency and bag life.

Most small- to medium-volume reverse-air collectors use a radial sweep arm on a central pivot to clean successive radial lines of filters as the arm is rotated continuously around a cylindrical housing. The air delivery can be controlled by a cam-and-valve mechanism, or electronic controls, to open the airflow while the arm is over each line of filters. Then the valve closes off the flow and permits the volume in the arm to serve as an air pressure reservoir. The cleaning air can be designed to flow continuously and enter the filters as the rotating arm sweeps over the line of openings.

Larger reverse-air collectors are generally divided into modules or sections to allow for cleaning selected sections off-line. This can be done sequentially, or by timers and valves that allow cleaning air into one module at a time. Butterfly dampers or poppet valves are used to open and close the airstream inlet, and/or outlet, to each module automatically as controlled by the cleaning cycle timer. Sizing of modules should be planned so that adequate filter area is maintained on-line during the cleaning cycles.

Each cycle will generally have a pause time, then the cleaning period, and then another quiet or pause time to allow the dust that was cleaned off to settle out. The cleaning circuit should have a photohelic control switch added to allow selection of the differential pressure level across the filters that initiates cleaning for the most efficient dust cake and filtration levels.

Test data support the benefit of building up and maintaining an optimum depth of dust cake on the filters in a reverse-air collector to produce the following:[8]

- A higher filtration efficiency since the dust must pass through this added depth of dust cake
- A pressure drop that tends to be lower and to level out at an acceptable differential pressure for a longer period
- Cleaning that should remove just enough dust cake to regain close to the original pressure drop, yet maintain good collection efficiency through a residual level of dust cake

Reverse-air primary collectors operate at a filtration velocity as specified for low-pressure cleaning in Table 2.2, usually in the 2.0- to 5.0-ft/min range. Large powerhouse or utility collectors use bag filters with a diameter of approximately 12 in, a length of 30 to 40 ft, and filtration velocities of 1.75 to 2.5 ft/min on coal fly ash from various mine sources. As mentioned earlier, secondary filter collectors have been operated at filtration velocities as high as 15 ft/min.

Large-particle dusts such as wood chips and heavy sawdust have also often allowed collectors to use a V_f of around 15 ft/min. These have typically used round or oval filter bags with a diameter of from 4.5 to 6.25 in, or equivalent oval sizes, and a length of 8 to 14 ft. Typical banks of multiple reverse-air collectors are shown in Figs. 1.5 and 1.6.

Pulse jet

Pulse-jet cleaning typically uses compressed air or gas, usually at pressures of 70 to 100 lb/in^2 gauge, directed through fixed blowpipes. Nozzles or drilled holes in the blowpipe are centered over each filter to deliver a narrow conical blast into the filter. Typically a diaphragm valve opens for a pulse of 50 to 100 milliseconds (ms) to inflate the filter and deliver a reverse-air blast that breaks up and cleans the filter cake off the dusty side of the bag (Fig. 3.2).

In pulse-jet cleaning the sharp pulse of air pressure breaks up the layer of dust cake by inflating the bag and dislodges the dust layer from the filter surface better than other methods. However, there is the precaution that the pressure can be too high for the filter media. So the pressure must be regulated to prevent premature damage or a reduced service life, especially with fragile media or surface coats such as Teflon membranes. A very fine particle size, moisture, or other dust cake difficulty may make it tempting to increase the pressure to clean better. But excessive pressure may damage fragile filter media, the sewn seams, or reduce the useful service life and so must be used with caution.

There are other routes to improve cleaning and filtration rather than just raising the cleaning pressure. The process may allow for the reduction of the inlet volume, dust loading, or moisture content, but most of these are necessary operating levels and must be maintained except for short times. Better alternatives in many cases, using a safe pressure, may be any of the following:

- Increasing the frequency of cleaning
- Using multiple pulses at short intervals
- Setting the photohelic controller to initiate cleaning at a lower differential pressure

Figure 3.2 Pulse-jet–cleaned filter collector, top door access. (*Courtesy of Staclean Diffuser Co.*)

- Using improved diaphragm valves (see below)
- Using off-line cleaning
- Using a filter aid powder injection or process dust cake buildup

By itself, pulse-jet cleaning of filters is the most effective method yet devised and is continually being improved. One improvement is the reduced cleaning air pressure of 15 to 35 lb/in^2 gauge under development use in IFB (industrial furnace and boiler) and utility station collectors.

New developments. A significant improvement is a controlled air consumption diaphragm valve that delivers a sharper, or shorter timed pulse for delivering pulse-jet cleaning air to the blowpipes. The air is delivered in a pulse that is briefer both in rise and cutoff time than with present valves. By its shorter pulse, this valve reduces the consumption of compressed air or gas for an equally effective cleaning pulse.[1] Due to improved flow characteristics it also reduces the required air supply pressure for an equal flow rate. The benefits are

- Improved cleaning effectiveness
- Lower pressure imposed on the filters
- Lower operating cost by reduction of compressed air consumption
- Reduced size and cost investment in air compressor capacity and annual energy costs

Diaphragm control valves are primarily made of cast aluminum with Buna or Viton diaphragms, depending on the service temperature, and are used with solenoid valves controlled by a timer. Corrosive environments may require type 316 stainless-steel valves with Teflon seals.

More sophisticated control circuits and programmable logic controllers (PLC) are being used to include all collector cleaning controls in one enclosure. They also provide for the selection of interlocking with other system equipment, local manual control of collector and accessory functions, or automatic and remote control.

Sensor transmitters are used to provide signals, such as 4 to 20 mA, for remote or centralized measurement and control. PLC or computer control of an area or process is thus permitted and is being used more frequently as plant personnel become trained in PLC programming and computer control for optimized operations. These control systems can be easily modified or expanded by reprogramming.

Careful design and selection of a pulse-jet cleaning system includes optimizing the method for delivery of air or gas into the filters. Size and number of blowholes should be matched to the valve and blowpipe size to balance flow and pressure delivery. Blowhole shape and pipe wall thickness should be selected using manufacturer's guidelines for the best performance, and the use of outlet nozzles on the blowpipe should be considered for their delivery benefits.

Cleaning air supply header capacity must be adequate for the number of blowpipes and valves. Provision for dry and clean air or gas is critical in many cases, even if just an automatic moisture drain is provided on the header. Moisture fed to the filters by the cleaning air is just as much a problem as moisture in the process stream and is often overlooked.

An obvious need is an adequate supply of air or gas at a regulated pressure. Driers, filters, drains, pressure regulators, heaters, and insulated enclosures must be considered on the cleaning air system. So rough-and-ready blowpipes on an inadequate supply line are being supplanted by systems designed for optimum cleaning with flows calculated for balanced delivery of pulse-jet air.

Telltale indicators at cleaning. On-off cleaning when the dust conveying blower or fan is shut down not only benefits all other cleaning methods to some extent, it also provides a telltale indicator of the collector's cleaning performance. If a great quantity of dust drops down into the hopper at shutdown, reentrainment during normal operation is indicated and may be a problem. This dropout can be detected audibly or visually if the discharge feeder is left running for a few minutes and continues to discharge dust, which is generally a good practice anyway to be sure that any dropout is removed.

If the collector has been running with a high delta P, or differential pressure, then dust dropout at shutdown probably indicates operation right on the edge of trouble. Some possible causes are air volume too high and thus passage (can) velocity also too high, too small a collector, filters spaced too close together, poor cleaning action, or difficult dust properties such as moisture or a large percentage count of fine particles below 0.5 μ.

However, some dropout after shutdown is normal in most cases and especially with sawdust, coarse plastic powders, and other light-density large-particle-size dusts. Experience and tests show that many such light-density large-particle products can be readily cleaned at high can and filtration velocities.

For dusts that are difficult to clean, off-line cleaning may become necessary with the collector split into modular sections. Planned on-off cleaning is a good consideration for chemicals and powders with a large percentage of the population count in the submicron range, moist and agglomerating products, or combinations of these. This off-line cleaning allows a long dropout time while the collector is in the quiet, or no upward flow, mode.

Economics usually justify off-line cleaning on larger reverse-air collectors on powerhouses, industrial boilers, and cement kilns. In these cases, a modular design is used to permit one section to be cleaned at a time, without a drastic reduction in filter area or increase in filtration velocity. Modules can also be used on smaller volumes to alternate operation between two equal size collectors, thus providing frequent off-line cleaning.

Can (upward passage) velocity. When the dusty airstream can or passage flow direction is counter to the dust discharge gravity flow, it may limit dust removal by any cleaning system. If an upward dusty flow lifts the dust particles in the airstream removed during cleaning cycles, they will be reentrained rather than discharged. The finer the dust particle size, the lower the permissible can or passage velocity between the filters.[4]

If the design can velocity is too low, the initial collector size and investment may be excessive for the process volume. Thus the value of past experience or tests will prove beneficial in selecting the optimum passage or can velocity. (See Fig. 3.3.)

$$\text{Can velocity } V_c = \frac{\text{airstream volume}}{(\text{housing cross-sectional area} - \text{filter cross-sectional area})} \tag{3.2}$$

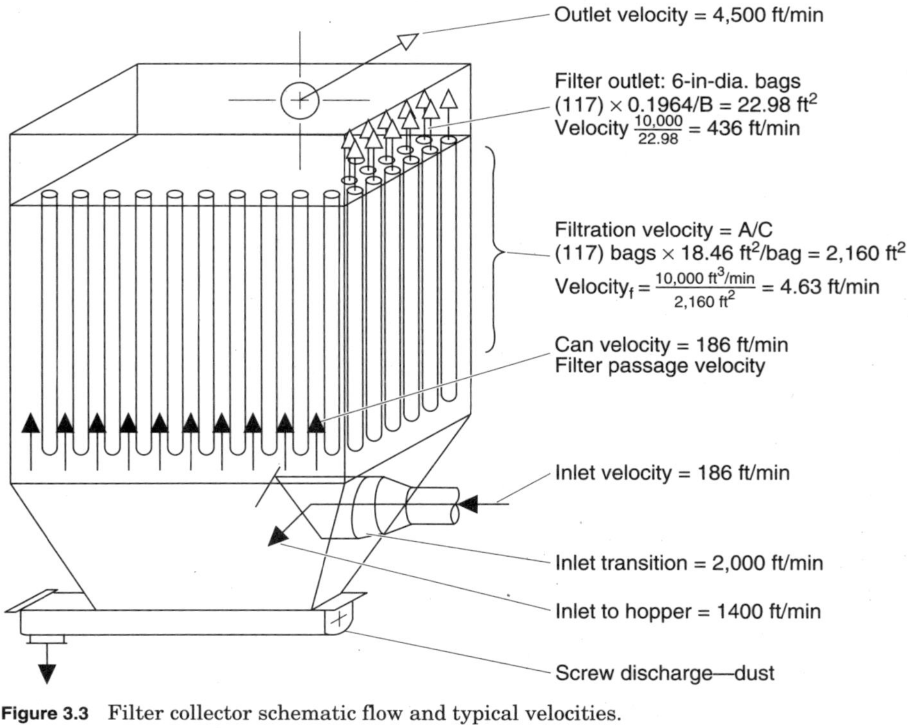

Figure 3.3 Filter collector schematic flow and typical velocities.

If a typical collector with a median particle size count of 5 to 10 μ has a footprint of 8 ft × 12 ft = 96 ft², then a fairly typical can velocity will be produced. If it has a median particle size count of 1 to 5 μ (fines), then it will probably have a footprint of 12 ft × 12 ft = 144 ft². But if the median particle size count is under 1 μ (ultrafines), then an even lower can is needed with a larger footprint area of say 12 ft × 18 ft = 216 ft²:

$$\text{Typical can velocity} = \frac{10{,}000\ \text{ft}^3/\text{min}}{96\ \text{ft}^2 - 41.64\ \text{ft}^2}$$

$$= \frac{10{,}000}{54.36\ \text{ft}^2}$$

$$= 184.0\ \text{ft/min}$$

$$\text{Fines can velocity} = \frac{10{,}000\ \text{ft}^3/\text{min}}{144\ \text{ft}^2 - 41.64\ \text{ft}^2}$$

$$= \frac{10{,}000}{103.4\ \text{ft}^2}$$

$$= 97.7\ \text{ft/min}$$

$$\text{Ultrafine can velocity} = \frac{10{,}000}{216\ \text{ft}^2 - 41.64\ \text{ft}^2}$$

$$= \frac{10{,}000}{174.4\ \text{ft}^2}$$

$$= 57.4\ \text{ft/min}$$

Can velocities may range as follows, for the dust particle sizes noted except that all unusual velocities should be evaluated in tests or judged by similar application experience:

- *Typical.* Need a can velocity from 150 to 200 ft/min.
- *Fines.* Need a can velocity of less than 150 down to as low as say 75 ft/min.
- *Ultrafines.* Need a can velocity of 75 down to 25 ft/min.
- *Off-line cleaning.* Will produce a can velocity of 0 ft/min.
- *Large particles* (10- to 100-μ or larger sizes of sawdust, plastic pellets, or similar low-density reasonably porous filter caking materials) *and heavy density particles* (60 lb/ft³ or heavier—cement, rock or metallic ores or others that will settle out without reentrain-

ment). May be able to use a can velocity of 200 to 250 ft/min. Above that is rarely satisfactory!

On-demand. On-demand cleaning systems should be used wherever possible to automatically maintain the most effective filter cake and filtration efficiency, as well as to achieve the most economical energy use in fan and cleaning power. The on-demand system uses a photohelic gauge controller to indicate the pressure drop across the filter media and two adjustable photohelic switches to control pressure settings.

A high differential pressure setting is used for starting the cleaning cycle, and a low-pressure setting is used for stopping the cycle. On-demand operation can reduce compressed air consumption, reduce wear on the filters, reduce fan power required, maintain an optimum dust cake, and increase collection efficiency, all for less than a $200 incremental cost.

3.4 Isolation Dampers or Valves

As mentioned above, one use for dampers or valves on input and output airstream connections to a collector is to provide for control of off-line cleaning. Another purpose is to provide for isolation of one module or collector for repair and maintenance, while the rest remain in operation. Another use might be for switching ductwork from one process input or output to another, thus allowing multiple use of collectors where compatible. Isolation dampers may not need to be totally leak-free for lower-temperature and nonhazardous environment maintenance work. Hazardous and high-temperature airstreams require adequate seals and leakage control for the exposure conditions.[4]

A further use of dampers is to bypass the collector during start-up or upset conditions. This would permit heating above a dew point condition prior to introducing product into the airstream or prior to introducing the airstream into the collector.

3.5 Materials, Coatings, and Manufacturing Quality

Materials

The material of construction normally considered first is that which would provide the housing and accessories at the lowest cost possible, so hot-rolled or A36 structural steel is the basic choice. The next step is the application of paint or a coating, as discussed below. A coating of ceramic, ultrahigh molecular weight (UHMW) plastic or a rubber lining, or a plasma- or spray-applied metal may be best for some abrasive, corrosive, friction, or temperature needs.

When the service becomes more severe, then the coated steel may be judged inadequate or sometimes may become more costly than an alternative base metal. Then the choice will turn to a more resistant material, such as 304 or 316 stainless steel, aluminum, Corten steel, fiberglass, polyvinylchloride (PVC), Inconel, Hastelloy, or titanium. Obviously those materials that cost several times more plus a premium to fabricate will only be selected where justified. The collection performance may not be initially affected by a poor choice of material, but a later failure will!

Paints

Paints and coatings are usually the first method considered to improve corrosion resistance of the basic low-carbon steel and can often be a successful lower cost alternative to higher cost materials of construction. The metal's surface preparation and roughness profile must be suitable for the paint properties and thickness. This is a common oversight or shortcut, but meeting the paint specifications for surface preparation will save cost in the long run.

Temperature of the operation is a basic hurdle for coatings. Many users will specify an epoxy or a urethane paint, for instance, for the gloss desired or corrosion protection. But if the operating temperature exceeds the coating's limit [usually 250°F (121°C)], then something in a silicone alkyd [up to 450°F (232°C)], or a silicone aluminum type [up to 1000°F (538°C)] must be evaluated. So consider the service temperature first. There are many paint choices up to 250°F, several good ones to 450°F, but only a few suitable for 1000°F temperatures.

There are new coatings being developed constantly, just as in most fields. A recent silicone aluminum is available that will provide resistance in the 400 to 1000°F (204° to 538°C) range with peaks to 1200°F (649°C). Another is a combination of ceramic frits and silicone resins in a coating capable of withstanding temperatures up to 1400°F, and the life expectancy actually increases as the service temperature increases up to 1400°F (760°C). While these new coatings may withstand some thermal shock, corrosion, chemical attack, and weathering, the resistance needs to be confirmed for any exposure.

Thin Teflon coatings with continuous $\frac{1}{16}$- to $\frac{1}{8}$-in thickness may be a good alternative. Sheets of Teflon, UHMW, or high density polyethylene (HDPE) can be a good low-friction surface and good in abrasive wear service if the joints in the sheets do not present a problem. Ceramic coatings can be applied in tiles, or in a continuous coating, for high-temperature and abrasive wear service.

When the corrosive, temperature, and moisture conditions get severe, the coating choices are limited. Coating suppliers can and should recommend a choice from their experience or from preliminary tests. But any collector supplier takes a needless risk by accepting full responsibility for a coating system to meet an unfamiliar or variable exposure. The ultimate users would do better in the long run by making their own selection or approving a supplier's recommendation, based upon valid experience or tests.

The coating selection should best suit the user's known exposure conditions or be based on tests for a new process exposure. Thus all parties including the paint manufacturers would minimize failures by a cooperative determination of the most effective coating. This beats trying to correct a major problem in the field and trying to determine whose fault it is after the fact.

A new process, product, or service may present uncertain exposure conditions even to the final user. All parties should then work together to select a coating and materials of construction that will best meet the service expected. That should achieve the best solution for all. To do otherwise is to invite disaster!

Quality

It is essential that the fabrication and manufacturing meet the job requirements and specifications. Structural integrity, interior finish, air leakage, cleanability, operating performance, and service life are affected by forming, welding, fastening and sealing, finishing, coating application, and fit of components. Any of these can be a critical consideration, such as for Food and Drug Administration (FDA) approval.

Forming must follow the design and specifications accurately for top results. Round collectors need to have the shell smoothly formed for strength and for many of the reasons mentioned above. With good suppliers this quality is usually a matter of course.

Welding should always meet good practice for strength and dust-tight construction and must meet ASME code requirements if necessary for pressure above 15 lb/in^2. Where corrosion is a concern, the welds should not cause embrittlement nor contamination and should be ground and cleaned to suit the job specifications.

Finishing to a suitable polish is necessary to meet FDA or process cleanliness when required. Sometimes a smooth finish can simply be to improve gravity flow of material out of a hopper. Another reason to select a fine polish can be to prevent contamination when switching to different input materials. The FDA regulations may dictate an inspection and certification of finish for processing food. So consider the added cost if this is necessary.

A good fit of components is expected but is not always received. The tolerances required are usually not tight, just a good fit for field assembly and for leakproof joints. Doors, access ports, and all operating parts should fit as required. Ensuring identical bolt patterns for flanges and mating accessories is important and should be straightforward. A good practice is to check whether such needs are specified in the beginning.

Fastening and sealing are the final means of closing up the collector for successful service. If the fit of components and other quality factors have been satisfied, then sealing should not be a problem. Remember the temperature expected and be sure all seal gaskets, caulk, parts, and accessories are specified to meet that expected level.

A critical fit and fastening area of concern is the filter attachment at the tube sheet. This should be dust-tight and easily maintained throughout the service life, since this is where the dusty air is separated from the clean air. The tightest seal on a bag filter is usually maintained with a snap band top, which has a spring band sewn in the top of the filter felt or cloth. This band or ring fits very snugly in an accurately formed tube sheet hole and is nearly foolproof in maintaining a tight seal. Top removal permits bags to be changed in the clean-air plenum, thus minimizing work in the dirty-air plenum.

The various clamps that are used on bottom removal bags must be carefully applied and tightened to secure a good seal. Bag changing must be done in the process dust or dirty-air plenum, which is a problem in some collectors.

Final approval

Quality can be backed up by several means. Field inspection during fabrication helps but is an expensive process. Besides, the ultimate quality achieved may only be determined during field erection and start-up operation. Dust collectors often require field assembly because of their size, and only on site do some component fit and operating performance problems become apparent.

A performance test requirement after installation is a good procedure. The test should be agreed upon to prove to both supplier and customer that actual performance meets that specified, within reasonable tolerances. There should be provision in the specifications for a second test in case corrections are needed, and the cost responsibilities should be spelled out.

A 5 or 10 percent retainage of total purchase price until final approval is often the best way to ensure quality. This puts teeth into ensuring that the customer specifications, equipment performance, and quality measurements are met. This retainage should be an advantage for both the user and for fabricators with good quality and design. Those suppliers that struggle to meet the specifications will incur extra costs and more difficulties at installation. Suppliers who maintain a good overall quality level find that it pays off in repeat business and referrals. In the words of W. E. Deming, “Quality does not cost, it pays.”

World-class quality

A recent quality improvement is the International Standards Organization (ISO) 9000 certification. This is a world-class quality certification that indicates approval of a product in all pertinent facets to ensure that the delivered performance and service are equal to that promised and usually better than products of nonapproved manufacturers. Some suppliers in the environmental and dust pollution field have already received certification for their products. *Chemical Engineering* magazine reported 11 new certifications of manufacturers for materials or equipment in the month of October 1993 alone.[4]

Certification should guarantee a consistently high quality level and should be increasingly available, probably at no cost increase in the long run. If the American manufacturing management had realized the need and achieved this level of quality 30 years ago, they could have outperformed the world continuously.

Housing strength

A collector enclosure must be strong enough to safely withstand the expected positive or negative pressure of the process. The most common exposure is a negative pressure or vacuum of 12 to 18 in WG, where the airstream is drawn through the filter. Housing strength is usually sufficient in a rectangular shape constructed of 12-gauge carbon or stainless steel with structural stiffeners on 32-in centers. A 10-gauge tube sheet with reinforcement on 30-in centers is usually adequate at this pressure to maintain a flat and level plane on which to mount the filter bags. Figure 3.4 shows the reinforcement spacing that should be used for various wall thicknesses and internal pressures.

Higher vacuums or pressures can be withstood by closer stiffener spacing, thicker steel for the skin, using a round cylindrical shape, or a combination of the above. A round design is inherently much stronger than a rectangular one due to its *hoop* strength, which more fully utilizes the strength of the metal skin in tension or in compression than does a flat surface. Figure 1.8 can be compared with Fig. 1.12, 3.2, or 3.5 for round versus rectangular designs.

A vacuum causes a more severe loading than internal pressure because of the complex forces caused by external atmospheric pressure trying to crush the shell inward. This crushing force causes around 5 times higher stress than an internal pressure. Internal pressure, on the other hand, exerts a uniform loading on the skin similar to that of a balloon skin, which utilizes the full strength of the wall thickness or cross-sectional area.

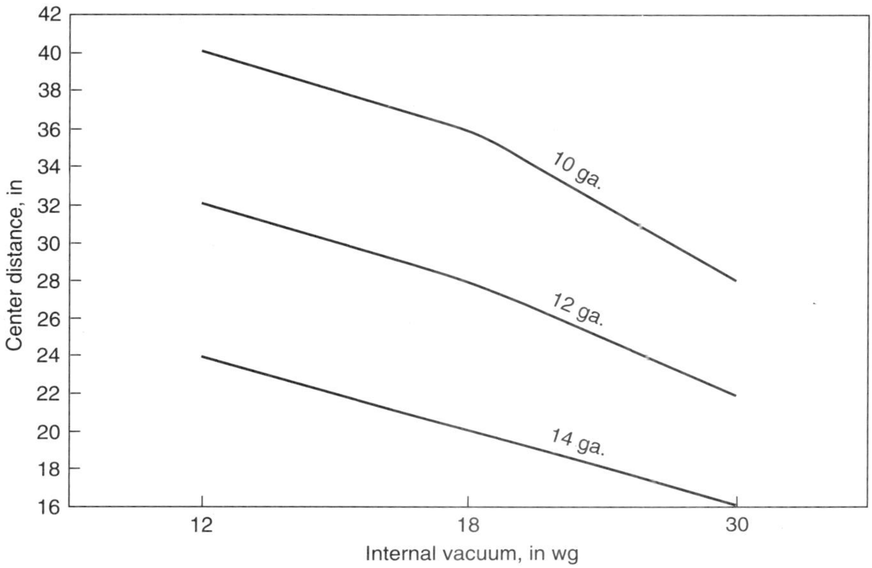

Figure 3.4 Flat wall reinforcing centers. Reinforcement spacing versus wall thickness and internal pressure.

Submarines use a cylindrical shape for their hull for maximum strength to resist the crushing pressure of the sea A streamlined outer shell or fairing is used to reduce drag, but it is not subjected to the full pressure that the hull experiences. The use of round collector design becomes economically imperative at negative pressures above about 30 in WG. It is advisable to have the supplier specify the housing strength rating, usually in operating pressure or vacuum in inches water gauge.

Structural strength

It is also advisable to have the supplier certify the structural strength of the collector and supporting structure, especially for outdoor exposure to wind, snow, and seismic loads. Building codes for most localities around the world are available and should be verified for structural design. Suppliers can design and certify that the specified loads and codes are met. If a professional engineer's stamp of certification is necessary, or a special code requirement or load must be met, then additional cost will probably be involved. Many user companies have set their own strength requirements based on past experience, which may vary from normal codes. These requirements must be specified for correct supplier design.

Be realistic in specifying the maximum expected amount or weight of dust material to be retained in the collector. Some operations will require that the weight of a full hopper of dust must be supported, but this may well add to the cost. Consider also that a load in the hopper at start-up will put a 100 percent loading on the discharge equipment, such as a screw conveyor, and the drive and strength of construction will also have to be beefed up (see Chap. 6).

The total amount of fill material in a hopper may not reach 100 percent if the dusty air inlet is in the hopper since the inlet will become choked off before the hopper can be filled. This maximum fill level would probably, on the average, reach to the horizontal centerline of the inlet when blown in through the hopper inlet.

Shipping

Truck shipping has width limits of 14-ft maximum in most states. Some states and countries will permit wider loads, but movement will then be limited when passing through states with the 14-ft maximum. Transportation by barge or ship would allow much wider sizes, but access to and from the plants and shipping docks restricts water movement to selected cases. Railroad shipments may be as wide as 17 ft. Air shipments will vary due to the carrier and countries as well as the aircraft available. (See case study 8 for alternatives.) Costs and transit time must be considered for any alternatives other than truck shipment.

Thus all large shipment sizes should be verified before a commitment is made, since the final limits may vary for the specific locations, modes, and routes. Total weights have not usually been a problem since most collectors do not reach the maximum weight limits.

As pointed out in one of the case studies discussed below, modular construction of collectors may offer an advantage in shipping. By allowing complete modules to be shipped assembled, both shipping and installation costs may be reduced versus one large collector. Where a complete collector may need to be broken down for shipment and then assembled for installation, the total cost may be reduced by modular construction.

3.6 Case Studies

Case Study 1: SO_x-NO_x-Rox-Box (SNRB) Patented by Babcock and Wilcox, SNRB is a filter collector for combined SO_x, NO_x, and dust particulate (Rox) air pollution control all in one enclosure (Box). The technology has been developed so that all three pollutants are removed from the flue gas stream in one high-temperature pulse-jet bag house enclosure. The first 1000 h of operation have shown that a temperature of around 850°F (454°C) is optimum for the best combined emissions control.[5]

The high-temperature bag house is the heart of the SNRB process, and the use of 3M Nextel ceramic fiber filter bags is critical for successful filtration at this temperature. These bags are mounted and clamped on perforated metal cage-diffusers for better support of the relatively fragile ceramic fiber filters media than is provided by a wire cage. The perforated cage also serves as a diffuser for the pulse-jet cleaning airflow so that it is more uniformly distributed along the full length of the filter. This is similar to the Staclean Diffuser which was developed for the uniform cleaning advantage and was then found to allow reduced cleaning air pressure, improve effective filtration area and velocity, and extend the service life of filters.

The development SNRB pulse-jet collector is a six-module design. Each module contains six rows of seven bags. Filtration velocity (air-to-cloth ratio) was about 4.0 ft/min. Filter cleaning was activated at a filter pressure drop of 13.5 in WG for on-line cleaning. Obviously the six-module collector could be cleaned with one module off-line at a time for better dust discharge from the filter passages if excessive can velocity becomes a problem.

Environmental Protection Agency method 5 sampling was used to measure the particulate emissions (Rox) at the bag house outlet. These emissions were consistently below the NSPS standard and project goal of 0.03 lb/million British thermal units (Btu). Most were less than half the NSPS standard and the average with hydrated lime injection for SO_2 control was about 0.012 lb/million Btu. Greater than 80 percent removal of SO_2 exceeded the goal of 70 percent. Removal of NO_x greater than the 90 percent goal was also achieved at the operating temperature of 850°F (454°C) with a catalyst inside the cage.

So this initial operation has shown that all three pollutants can be removed in one collector housing (Box) at better than the project goals and can be done so at compatible operating conditions up to temperatures of 1000°F (540°C). There should be great promise in the application of this development in filter collection for chemical, cement, and other industrial furnaces and boilers (IFB) and power generating stations.[6]

Case Study 2: Asphalt Plant Aggregate Contaminated aggregate and dust after being dried was collected in an existing filter collector. Plant capacity was limited by excessive emissions at higher throughputs, and fine-particle penetration into the filter bags shortened their useful life. Also, the pressure differential across the filters climbed too high when capacity approached design rates. But the plant needed to expand production capacity to meet market demand.

Tests were conducted on inlet, hopper, and bag dust samples, and filter bags removed from the existing collector were examined. The test results were summarized by volume (or weight) and by population or particle counts as shown in the following tables:

Volume	Median size, μ	% Below 1 μ
Inlet dust	57.44	0.045
Hopper dust	42.53	0.036
Bag dust	3.403	7.325

Counts	Median size, μ	% Below 1 μ
Inlet dust	0.614	72.50
Hopper dust	0.777	61.08
Bag dust	0.649	75.52

The large percentage difference between volume and particle count is due to the effect of size. Just one 50-μ particle has the same volume as one million 0.5-μ particles.

The significance of the tests and filter examination showed that while the inlet and hopper volume figures are in good agreement, the bag dust volume shows a substantial increase in fines volume and a drastic reduction in median-size particles. The accumulation level of fines at 7.325 percent by volume or weight on the filter bags versus the inlet and hopper levels points out the problem.

The can velocity must be too high and is causing reentrainment and excessive accumulation of fine particles on the filter bags. This accumulation will lead to fine-particle leakage through the filters and emission problems. Plus it probably increased pressure drop across the bags. Or as input is increased there will be more fines penetration, greater pressure drop, and thus more emission problems.

The solution was to design and install a second filter collector that was sized to handle the expansion plus some of the previous lack of capacity. Filtration velocity was reduced in the new design to 4.1 versus the existing 7.2 ft/min. The can velocity was lowered significantly. P-84 singed felt filter bags were selected in place of Nomex bags. (See Figs. 3.5 through 3.9.)

Figure 3.5 Pulse-jet filter collector, fan, and clean exhaust stack at an asphalt plant in full operation, with handling nuisance dust over silos to the left. (*Courtesy of Continental Paving Co.*)

Figure 3.6 Belt conveyor input to drier, with nuisance dust from handling visible, filter collector on left with no visible emission, at asphalt plant in full operation. (*Courtesy of Continental Paving Co.*)

Figure 3.7 Drier discharge and dirty-air inlet with enlarged settling section and cooling duct to pulse-jet collector hopper, at asphalt plant in full operation with no visible emission. (*Courtesy of Continental Paving Co.*)

Figure 3.8 Collector wall reinforcement, hopper with dirty-air inlet, discharge screw conveyor with optional start-up discharge to dump box, at an asphalt plant. (*Courtesy of Continental Paving Co.*)

Following successful start-up of the new collector, the original filter collector was overhauled and modified to meet the same operating specifications as the new unit. Rather than being limited in capacity by excess emissions as before, both units are now meeting the plant operating goals, as well as the pollution regulations. One filter collector alone can handle smaller-volume runs efficiently.

Collection efficiency and emission levels are now better than expected. When an environmental control board inspector arrived at the plant for an operating emissions test, he looked up at the stack and not seeing a visible plume of dust he told the supervisor they could start up any time. The supervisor told the inspector that they were already operating at a full production rate!

Case Study 3: Reverse-Air Fabric Filter for Fly Ash This filter collects fly ash from a boiler/power plant, has a volume of 2,200,000 ft^3/min, and an operating temperature of 300°F (150°C). A 40-module (40-compartment) design reduces pressure drop during cleaning of individual modules and provides a more uniform operating pressure on the induced draft fan.

Three hundred and twenty-four bags are used in each compartment. Since there are 40 compartments, there is a total of 12,960 filter bags. Each bag has a diameter of 12 in and a length of 32 ft 9 in, for an approximate cloth area of 97.0 ft^2 per bag after deducting for support rings. Thus the total filter area is about 1,257,100 ft^2. Filtration velocity is 2,200,000/1,257,100 = 1.75 ft/min. With one module being cleaned, the filtration velocity is increased to only 1.795 ft/min.

Bags are 14 ounces per square yard (oz/yd^2) fiberglass coated with acid-resistant polymer resin and a Teflon–silicon-graphite finish. Cleaning is reverse air

Figure 3.9 Clean-air outlet from new filter collector to fan inlet transition on parallel existing filter collector. (*Courtesy of Continental Paving Co.*)

at low pressure. Inlet fly ash dust loading is 1.0 to 3.4 gr/actual ft^3, or an average of 2.2 gr/ft^3. Emissions outlet is less than 0.007 gr/actual ft^3, and opacity is less than 20 percent. Calculated average efficiency is thus 99 68 percent [(2.2 − 0.007)/2.2 × 100%].

A precoat system is installed to apply fly ash as a precoat on new fiberglass filter bags. This precludes premature plugging of the fiber filters because of penetration of the fine fly ash particles. Overall collector size is 145 ft × 235 ft × 90 ft high. A PLC is used to optimize and monitor the operation of the system and can allow integration into the main boiler control panel.

A penthouse protects maintenance workers during bag changes or other work as well as the equipment such as module isolation dampers from the weather. There is provision for a preheat system if necessary in the future; and a vacuum cleaning system is permanently installed for cleanup of dust during maintenance (see Fig.3.10).

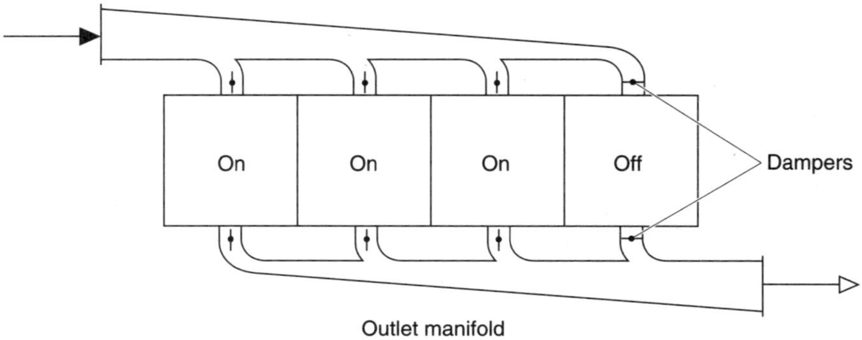

Figure 3.10 Modular filter collector for off-line cleaning and service.

Case Study 4: Filter Bags Penetrated by Ultrafine Dust Sixteen-oz/yd^2 polyester needled felt singed filter bags were being penetrated by fine furnace coke dust with a sulfate content in a chemical plant pulse-jet collector. Laboratory tests revealed full-depth dust penetration through the felt filter media. This is a typical case with many ultrafine chemicals and powders.

The embedded dust in the filter media was analyzed and found to have 86.63 percent below 1-μ size by particulate count, with a median size of 0.539 μ. This dust does contain an extremely high percentage of particles below 1 μ, and the collector does need a higher efficiency filter media. Several possibilities for consideration are

1. Heavier polyester felt, but not singed
2. Teflon coat or membrane on the felt polyester
3. Duo-density felt
4. P-84–type felt
5. Improvement of filter cleaning methods

An 18-oz/yd^2 polyester nonsinged felt was selected after evaluation and has been used successfully for several years now. While singeing usually helps dust release from the surface, it does form tiny nodules of melted fibers on the surface which effectively reduce the open area and increases the filtration velocity. A retest of the dust showed it had 47 percent by particle count below 0.5-μ size and a bulk density of about 35 lb/ft^3.

The can velocity was 199 ft/min, which is high for collection of a dust with this large quantity of fine-particle sizes. Future collectors for this service should consider a design with lower can velocity to reduce the concentration of fines retained on the filter surfaces. The sewn seams on the filter bags were found to be the only current source of fine-dust leakage. Future filter bag trials should be conducted with fused seam construction, since that eliminates the needle holes through the felt.

The pulse-jet filter cleaning air pressure was reduced from 90 down to 70 lb/in^2 gauge, and frequency or time between pulsing was lengthened. The furnace dust was still adequately removed, and a slight lowering of pressure drop across the filter bags has been achieved. Apparently the retention of a better filter cake on the bags for a longer time was the reason for improvement. It is difficult to maintain an adequate filter cake so a good precoat, a moderate cleaning pressure, and a long cycle time are beneficial.

Case Study 5: High Differential Pressure Drop on a Pulse-Jet Baghouse A filter bag was removed and tested from a collector on the exhaust stream of a chemical plant incinerator generating fine silica particulates and fumes. The collector was producing high pressure drops, and the permeability, or flow, regain after pulse cleaning was low.

Test results showed very poor release properties of the fine dust from the porometric membrane surface of the filter medium. This poor release produces low airflow, or permeability, through the filter. Dust loading on the filter and in the airstream quickly builds up and produces the high differential pressure across the filter. The flow did not improve above the initial cleaning level even after repeated cleaning pulses.

The dust on the filter surface contains a median count particle size of 0.326 μ with 90.20 percent of all particles under 1 μ. This is an extremely fine size for any collector. Test extractions also revealed a slight condensation contaminant in the dust, which combined with the extremely fine dust, produced the poor release properties, the low flow, and a high differential pressure.

Off-line cleaning had been found effective in trials, so the solution worked out was to build and install a second collector of the same size and install it so a two-way diverter valve could alternate the airflow between the collectors. An automatic actuator control was developed to switch the airstream to the cleaned bag house when the pressure drop reached a preset level in the operating collector. Preliminary operation showed good results with an off-line cleaning time of about 15 min, and on-line collection was effective for that time (see Fig. 3.11).

Case Study 6: Compact Hybrid Particulate Collector (COHPAC) The concept was to design a filter bag house for fine dust particulate collection after a electrostatic precipitator on a lignite-fired steam generating station. The precipitator removes a significant portion of the dust from the airstream, so the flue gas reaching the bag house has a greatly reduced dust load, and those particles may carry a residual electrostatic charge. As a result, trials have shown that the filter bag house can operate at very high filtration velocities of 10 to 20 ft/min without a significant increase in pressure drop or emissions.

The COHPAC bag house can be much smaller in physical size than a conventional collector design using 2 to 6 ft/min filtration velocity. The COHPAC has better collection efficiency on fine particles than precipitators, which in addition to improving emission levels should help in the future capture of toxic gas. Twenty-foot-long filter bags are oval in cross section and arranged in a circular pattern in each module. A rotating cleaning arm delivers the cleaning pulse to each row of bags in sequence. The filter material is Ryton needled felt of 18 oz/yd^2 on Ryton scrim fabric.

After 3700 h and 6 months of operation, the COHPAC has met or exceeded the performance objectives. Opacity is less than 20 percent, differential pressure has been held below 6 in WG, mechanical draft losses through the collector are less than 2.5 in WG, and the operating filtration velocity has been maintained at 15.5 ft/min. Outlet emissions are 0.01 lb/MMBtu, and the low opacity has been achieved regardless of changes in fuel or flue gas conditions, such as sorbent addition to the flue gas.

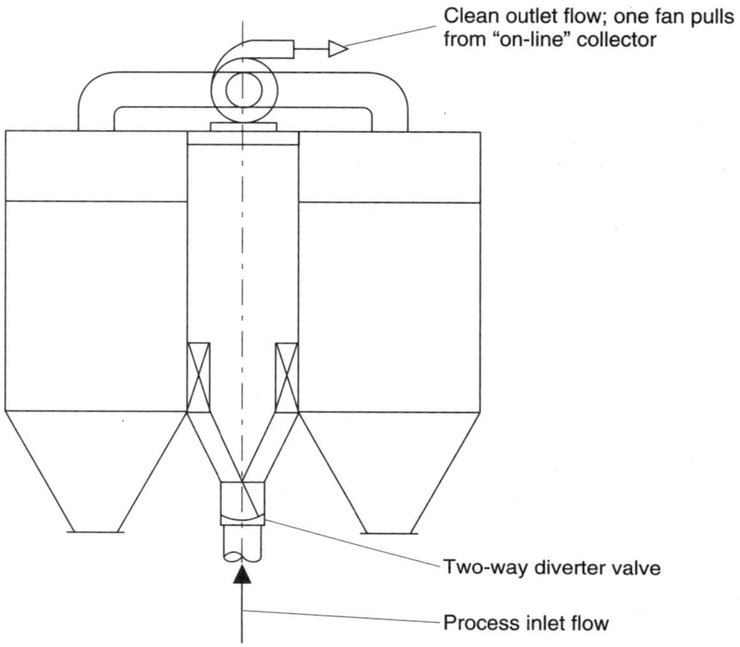

Figure 3.11 Two separate module filter collectors and flow diverter for alternate off-line cleaning. (Refer to Fig. 10.3 wiring diagram.)

The demonstration pulse-jet COHPAC uses a low-pressure high-volume cleaning system, with the 8- to 12-lb/in^2 gauge pressure supplied by a blower. The 145-MW scale bag house has eight compartments or modules and filters a flue gas volume of 735,000 actual ft^3/min at 385°F. Total cloth area of the bag house is 63,400 ft^2, but allowing for one module cleaning and one for maintenance, the on-line area is 47,550 ft^2. Thus the filtration velocity of (735,000 ft^3/min) /(47,550 ft^2 = 15.5 ft/min actual on-line .

Case Study 7: Pulse-Jet Secondary Collector A pulse-jet bag house was installed downstream of a reverse-air bag house instead of an electrostatic precipitator (ESP) for fly ash collection. Sodium sorbent was injected for SO_2 control before the pulse-jet bag house, which was operated at a high filtration velocity, around 15 ft/min, to collect the sorbent. This collection was conducted on an Electric Power Research Institute (EPRI) transportable pulse-jet (TPJ) fabric filter pulse-jet bag house.[8]

Most of the fly ash is removed in the upstream reverse-air collector, so that fly ash and sorbent can be collected and sold or disposed of separately. The downstream bag house can be operated at high filtration velocity like the COHPAC, by proper selection of the SO_2 sorbent size. The TPJ used an intermediate pressure of 20 to 30 lb/in^2 gauge for filter cleaning.

The low outlet emissions achieved with the COHPAC design were confirmed with the TPJ using the same 15.5-ft/min filtration velocity. The average pressure drop across the filters was 5.5 in WG with each bag being cleaned an average of 2.8 times an hour. Inlet particulate concentration was 160 mg/N · m^3 and outlet emissions were 1.7 mg/N^3 as measured by EPA method 5. Removal of SO_2 was only a low 36 percent due to low flue gas temperatures of less than 300°F (150°C), but should be improved at higher temperatures. (See case study 1.)

Case Study 8: Standardized Modular Design System By designing a standard module that is a complete collector and making it economical to build, these standard building blocks can be combined to make up a complete filter collector system. The total airstream volume needed can be handled by including as many modules as are necessary (see Fig. 3.12 and Ref. 9).

This modular design approach with sizes up to 5950 ft^2 of filter area is now available, in footprint sizes from 5.5 × 5.5 ft up to 12 × 12 ft, or plan areas of 30.25 to 144.0 ft^2. So two modules provide 11,900 ft^2 of filter area, four provide 23,800 ft^2, eight provide 47,600 ft^2, and so on. Another feature is the connectability of the modular building blocks into an integrated whole. Shipping by truck should be greatly facilitated with a 12 W × 12 H × 36 ft to 12 × 12 × 40 ft load that clears all U.S. state size limits.

Inlet and outlet manifolds are available for connecting the modules' airstreams together to provide a single connection from the integrated collector to plant process ductwork for the total input of dusty and outlet of clean air. An inlet damper valve can be closed for maintenance or off-line cleaning of each module, also an outlet poppet valve can do the same. The modules can be located in banks on each side of the central manifolds, and the access walkway can be placed on top of the manifolds. Actuators for the outlet damper valves, as well as the cleaning system's pulse-jet valves and air supply headers, can be located on top of each module for easy access.

If the blowpipe and pulse-jet cleaning assembly are affixed to the top panel, then a hoist can lift off this entire top panel assembly for filter bag, cage, and diffuser service. An auxiliary penthouse weather enclosure with an overhead manual chain hoist and monorail can be located over the entire collector area to permit easy maintenance access.

Several inlet configurations are possible for even distribution of the airstream inside each collector and up through the filter bags. A set of horizontal ladder vane baffles stretched full width across the hopper and dividing the airstream in sequence would be one good solution.

Operational advantages of the modular design would be their easy use in off-line cleaning of one module at a time, and another advantage is easy expansion when needed. While any filter collector is readily adapted to injection of filter aid coatings, dry sorbent lime, or other powders, this building block design with the common input manifold is certainly suitable to such injection.

A further design consideration for current and future use would be bag-and-cage attachment to the tube sheet in an easy fashion, such as with snap band bags; dependable fastening for vertically plumb positioning of the bag and cage; variation in bag spacing for different can velocities at the design stage; and a bag-and-cage mount that will permit the use of sleeve-type filter bags such as fiberglass, Teflon, ceramic, or metal fabrics for higher-temperature operations. Bag-and-cage mounts should also permit the use of catalyst cylinders inside the cage for NO_x or other toxic gas reduction or neutralization.

While this idealized modular design may not suit every application, especially not a high-pressure or vacuum service that would most economically be

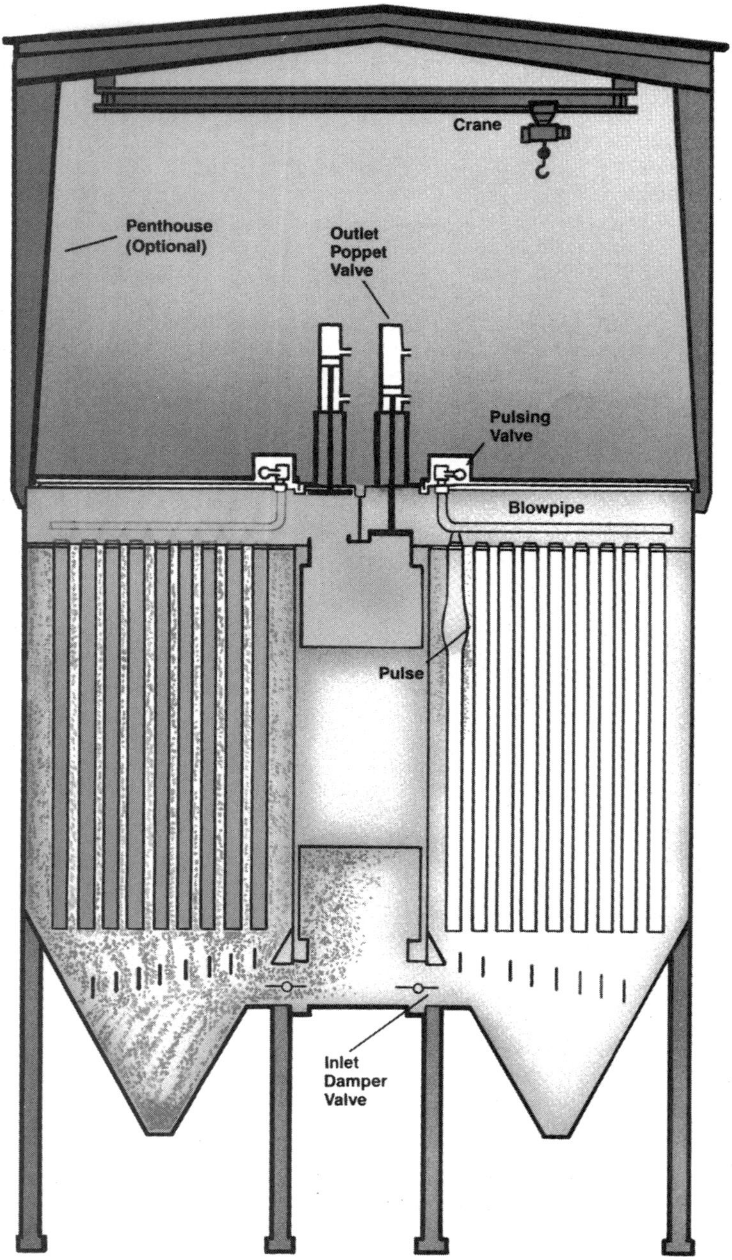

Figure 3.12 Standardized modular filter collector design. (*Courtesy of United McGill Corp.*)

served by a cylindrical design for shell strength, it should be evaluated for benefits in cost and accommodating many of the newer technologies applicable to filter dust collectors. Dust collection is here for the long haul, and advantages of filter separation are great enough to exploit every economic and design benefit.

Case Study 9: Cartridge Filter Collectors Folding the filter media into accordion pleats and forming the folded filter into a cylinder puts a lot more area into a given space. This is the premise of the cartridge filter that has carved a wide niche for itself in the last decade. Paper or cellulose filters were the first important media in cartridges and still are in low-temperature and low-cost applications, especially in air intake uses. Then felt materials were adapted for their advantages in temperature and collection efficiency.

Recently the effort to cram as many folds or pleats into each cartridge for maximum filter area has subsided, as the problems of dust caking in the narrow pleat opening spaces has become critical in many applications. While from 250 to 320 pleats was the norm a few years ago, design finesse has turned more toward reducing the number down to 150 or even 90 per cartridge. This reduces the total filter area available but has solved some difficult filtration by providing more effective cleaning release of dust from deep inside the pleats.

The space advantage of cartridge collectors must be taken into account where possible, with reduced height and footprint area as two benefits. A competitive cost structure has been achieved by the volume of business and by the marketing of standard sizes which gains some mass production benefits. Such standardized cartridge designs would be readily adapted to the modular building block approach to handling large airstream volumes (see Fig. 3.13).[10]

Case Study 10: Cement Nuisance Dust and Secondary Filters In a cement plant area where nuisance dust from handling, conveyor transfers, bagging operations, and welding fumes were objectionable, a special collector was needed to clean and recycle or return the air back into the same plant area. Workplace ventilation required a total airstream volume of 22,000 ft^3/min.

Since the dust loading would be very light, a high filtration velocity was selected of 24.3 ft/min. Sixteen-ounce singed polyester felt filter bags were used for the primary collection. After determining that off-line cleaning could be used approximately every 2 h at break periods, even the can velocity was set very high, at 875 ft/min. The operation has been successful with these high velocities because of the very light loading and frequent off-line cleaning.

Secondary filters of activated carbon were mounted in one entire vertical wall of the filter collector clean-air plenum to remove any objectionable odors from the clean airstream. Seventy-two square feet of 1-in-thick carbon filters would see 305-ft/min filtration velocity to handle the 22,000 ft^3/min. Expected pressure drop across the carbon filters would be 0.25 initially and 1.0 in WG finally. The operation has been successful in cleaning up odors, vapor, and nuisance dust and meets the necessary workplace standards (see Fig. 3.14).

Case Study 11: Dual Filter Collectors, Secondary, and Accessories A ceiling-tile plant handling a very light-density 8- to 12-lb/ft^3 fibrous trim and cutoff waste, installed two 15,000 filter collectors to handle the 30,000-ft^3/min volume

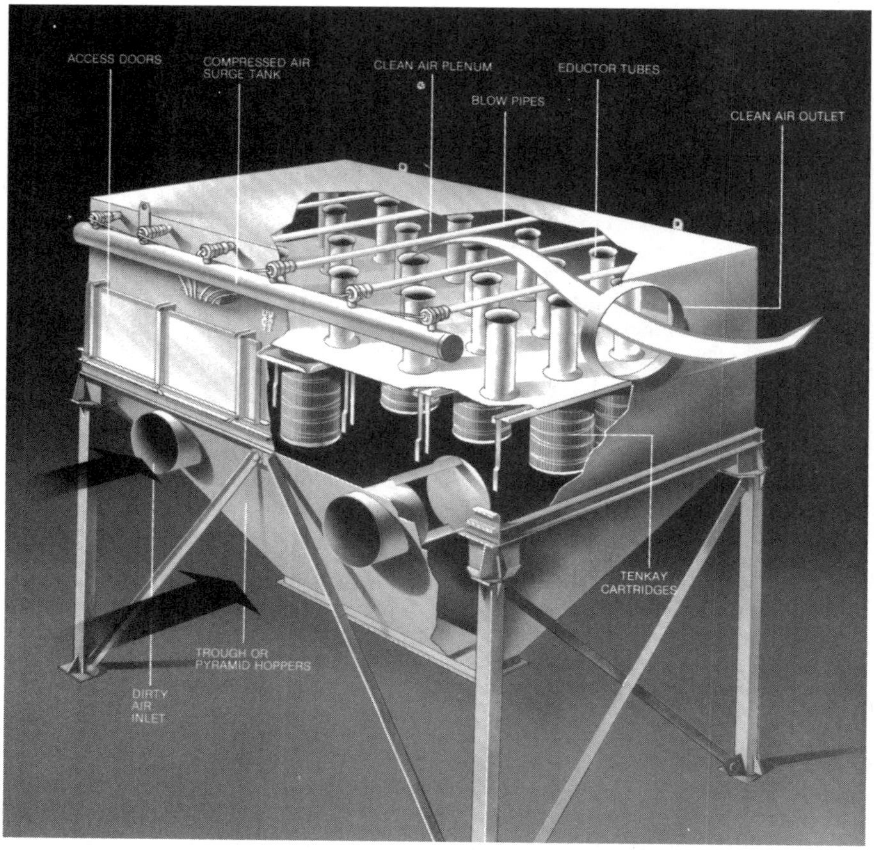

Figure 3.13 Cartridge filter collector layout and elevation. (*Courtesy of Farr Co.*)

required. Two collectors were specified because of space restrictions on the roof location and to make use of existing steel support structure. Past experience indicated that a can velocity of 159 ft/min and a filtration velocity of 5.1 ft/min would be acceptable for this light product.

Secondary HEPA, high-efficiency, particulate arrestor type filters were added after the primary bag filter collector with 16-oz/yd^2 polyester felt bags with Goretex membrane. The HEPAs were used for final filtration redundancy to be sure of the 99.99 percent efficiency expected from the primary filters. Also a safety monitor pressure differential controller-gauge was used to measure and initiate action if the pressure drop across the HEPAs exceeded a safe set point. Thus the clean air could be safely returned inside the plant for makeup (see Fig. 3.15).

Explosion vents were used on the collector in accordance with National Fire Protection Association standard 68. Insulation of 3-in semi-rigid mineral wool

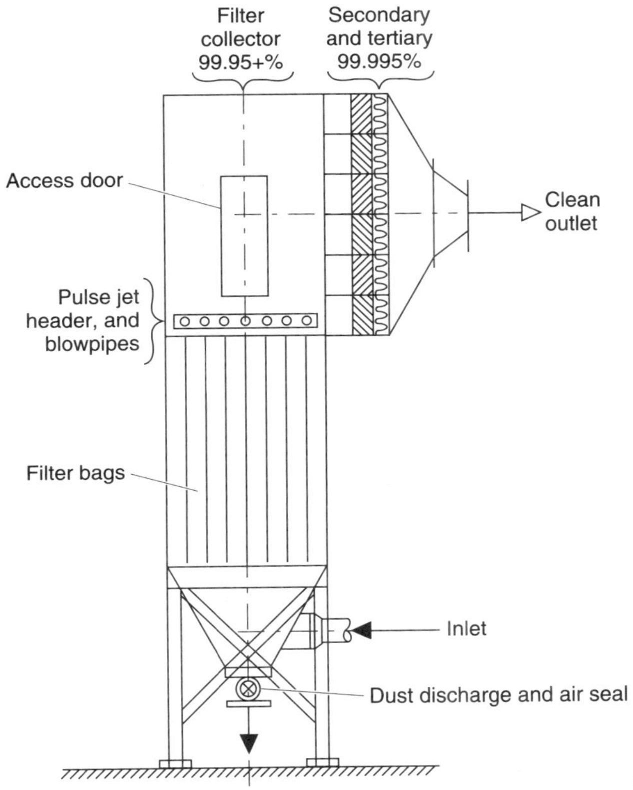

Figure 3.14 Integrated filter collector and secondary filters in the same housing.

and a 0.040-in-thick aluminum jacket were used to prevent condensation in the winter. A triboflow-type broken bag and emissions detector was used to measure and initiate action if dust particulate emissions exceeded the set point allowable. All these accessories are discussed in detail in Chap. 7 (See Fig. 1.5).

3.7 Conclusion

So the development of filter collectors goes on and seems to be accelerating as needed to meet ever-tighter emission limits. One of the keys to improved performance and collection efficiency has been the increased search for and use of evaluation methods to improve expected performance and then to maintain the design level in actual operation through regular logging of operating data and tests of performance levels.

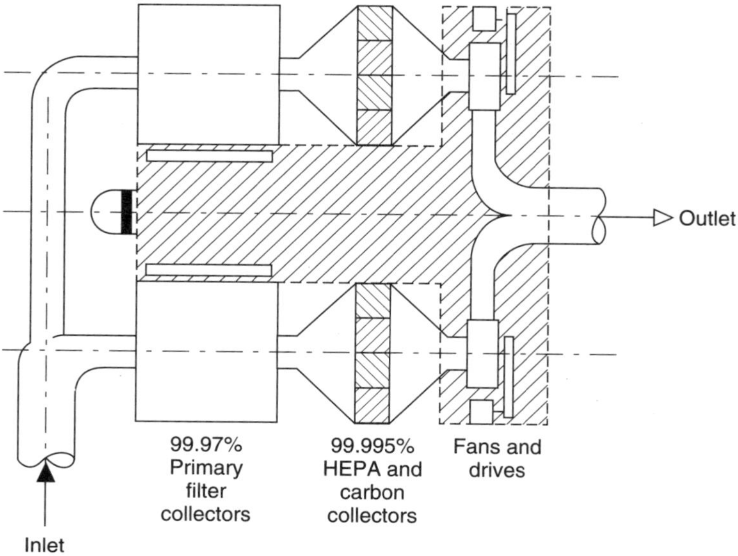

Figure 3.15 Dual filter collectors, dual secondary filters, and dual fans.

References

1. Goyen Valve Company, #FS-1092 "Dust Collector Valves, the FS Revolution," Lakewood, N.J. 08701, 12 pp.
2. U.S. EPA, "Electrostatic Augmentation of Fabric Filtration: Pulse-Jet Pilot Unit Experience," Nov. 1981, Research Triangle Park, NC 27711, p. 3.
3. *Chemical Engineering,* "Newsfront," Oct. 1993, p. 53.
4. United McGill Corp., "Fabric Filter Systems," "Gas Velocity and Distribution" section, May 1991, pp. 9, 12.
5. 3M, T. J. Gennrich, "Nextel High Temperature Filter Bags." St. Paul, MN 55144-1000.
6. J. M. Wilkinson, P. Chu, K. E. Redinger, Babcock and Wilcox, T. J. Gennrich and K. C. Hsieh, 3M, "5MWe SNRB Demonstration Project."
7. Environmental Consultant Company, Phoenix, AZ 85080. Winston Budrow, President.
8. EPRI-TPJ, Transportable Pulse Jet Fabric Filter Baghouse, Palo Alto, CA 94303.
9. United McGill Corp., "Fabric Filter Systems, Collectors for High-Efficiency Air Pollution Control," Columbus, OH, 1991.
10. "Dust Collection Systems," catalog, Farr Company, Los Angeles, CA 90009, 1993, Aspirated Cartridge Collectors Section.

Chapter

4

Optimum Sizing of Filter Collectors

4.1 Optimum Sizing and Design of Filter Collectors

Sizing should be optimized as much as possible right from the initial design stage but will probably be accomplished in a gradual stepwise approach as different technologies and designs are considered and as an overall project system progresses. Consider the following performance and cost factors as alternative collector designs are developed.

- Design to minimize size, weight, and cost
- Optimize cost and performance combination
- Process changes for elimination or reduction of pollutants at the source
- Savings opportunities in conjunction with collection objectives, for example, by returning captured dust or cleaned air
- Filter technology and media selection
- Paint or coatings versus basic materials of construction
- Minimum flat plate areas for wind and seismic loading
- Energy cost optimizing
- Cartridge, bag, and other filter configurations
- Modular collectors and standardized building block combinations for total airstream volume
- Pollution control integration in the collector housing, such as SO_2, NO_x, or vapor and fume secondary filtering

- High-temperature filtration, now possible up to 1000°F (538°C) and above, with ceramic and metal filters
- Accessory optimizing—discharge feeders, fans and dampers, drives, controls, insulation, heaters, explosion vents, safety equipment, ductwork, emission monitors, and others
- Velocity limits and their performance benefits
  - Duct conveying velocity V_d
  - Inlet velocity V_i and entry loss reduction
  - Can or filter passage velocity V_c and cross-section uniformity
  - Filtration velocity V_f (*A*/*C* ratio)
  - Tube sheet or filter outlet velocity V_o, especially for cartridge and longer bag filters
  - Exhaust velocity V_x and transition losses
  - Round versus rectangular enclosure design, especially for higher negative or positive pressures

Emissions, efficiency, and environment

The measurable results of dust collection are the level of particulate emissions, the collection efficiency, and how well we are protecting or improving the environment. The additional benefits of pollution collection and control are increased product recovery or yield, improved control of product quality, and possible reduced energy use. Savings generated help underwrite present and future improvements in pollution control.

Measurement of collection effectiveness

This is usually stated either as the collection efficiency, or as an emissions dust load in grains per cubic foot. Often it is required that the stated efficiency be effective at a specified maximum particle size, or size range, usually stated in microns (one millionth of a meter, one thousandth of a millimeter, or approximately 1/25,000 of an inch). Efficiency and emissions are related as given in Eq. (3.1) which is repeated here:

$$\text{Collection efficiency} = \frac{\text{inlet} - \text{emission dust load}}{\text{inlet dust load} \times 100\%}$$

$$= \frac{\text{collected dust load}}{\text{inlet dust load} \times 100\%}$$

If an emissions limit of 0.01 gr/ft^3 is required for a 10,000-ft^3/min system with a 50-gr/ft^3 inlet loading, then Eq. (3.1) gives: efficiency = [(50 − 0.01)/50] × 100% = (49.99/50) × 100% = 99.98%. For an emissions limit of 0.005 gr/ft^3, Eq. (3.1) gives: efficiency = (49.995/50) × 100% = 99.99%.

Inlet grain loading

Converting the inlet grain loading per cubic foot into pounds per hour, (50 gr/ft^3 × 10,000 ft^3/min × 60 min/hr)/7000 gr/lb = 4285.71 lb/h. At 99.98 percent collection efficiency, the emissions would be reduced to no more than 4285.71 × (100 − 99.98)/100% = 0.857 lb/h. For an efficiency of 99.99 percent, the emission limit would be equal to 4285.71 × (100 − 99.99)/100% = 0.42857 lb/h.

Filter collector benefit

In either case the collector is filtering 2.143 tons/h, or 4286 lb/h, and emitting only 0.86 or 0.43 lb/h. A medium-efficiency cyclone used 20 years ago as a primary collector and operating at around 85 percent efficiency on the same process would have captured only 0.85 × 4285.71 = 3642.85 lb/h, and emitted 642.86 lb/h.

So a filter collector would capture 642.86 − 0.43 = 642.43 lb/h more than a cyclone and emit 1495 times less particulate matter to the atmosphere. That improvement in collection efficiency is capturing 642.43 lb/h × 8400 h/yr = 5,396,412 lb more each year!

Optimizing filter size

Use the optimum filter media and maximum reasonable velocity choices from experience, tests, or references, case studies, and suppliers' data. Case studies should help in understanding the different application options to best suit the project circumstances.

Let's start with a 10,000-ft^3/min filter collector for cement as discussed in earlier chapters. Equation (2.2) looked at filtration velocity for product collection, temperature above 225°F (107°C), median particle size of 50 μ, and dust load above 80 gr/ft^3. But first let's determine the optimum size with the ideal filtration velocity.

Assuming we have selected a filter-type collector and based on a preliminary filtration velocity V_f (or A/C) and a 6-in bag diameter, the following charts should help optimize the initial design parameters, which all affect cost:

1. Length of filter bags
2. Number of filter bags
3. Floor space or footprint required

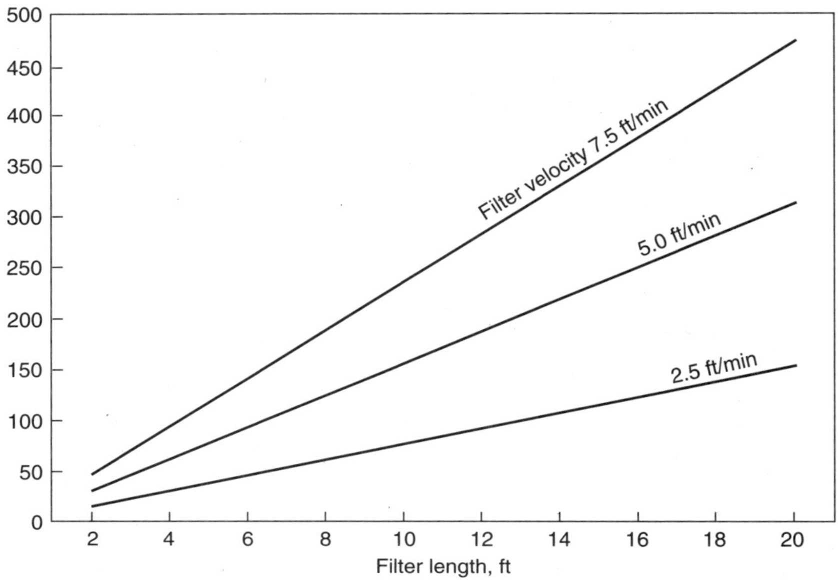

Figure 4.1 Optimizing can velocity versus filter bag length.

Figure 4.1 shows can velocity (V_c) versus filter bag length, for three filtration velocities (V_f) for the 10,000-ft^3/min volume, Q. Can velocity is measured as the airstream flow velocity between the filters. So with both a can velocity and the filtration velocity selected, the length of filter elements or bags can be determined.

Figure 4.2 shows the number of filter bags required for each filter bag length, at three filtration velocities for a 10,000-ft^3/min volume.

Figure 4.3 shows the floor area in square feet (footprint) versus filter bag length, at three filtration velocities for the 10,000-ft^3/min volume.

Note: All three figures (Figs. 4.1 through 4.3) are based on 6-in-diameter bags spaced on 9- × 10.5-in centers. Closer spacing is possible in many cases, but a good can velocity must be maintained. Wider spacing may be advantageous to reduce can velocity as needed while maintaining filter area and filtration velocity at the necessary design levels, but the footprint will increase. Also values for a different volume can be interpolated from these figures for 10,000 ft^3/min, i.e., 30,000 ft^3/min requires 3 times the number of bags or floor space as a 10,000-ft^3/min collector. But a 5000 ft^3/min collector needs only half as many. So these chart values can be used as a quick reference for any volume, and they should agree well with those calculated item by item, which should be particularly helpful in preliminary planning.

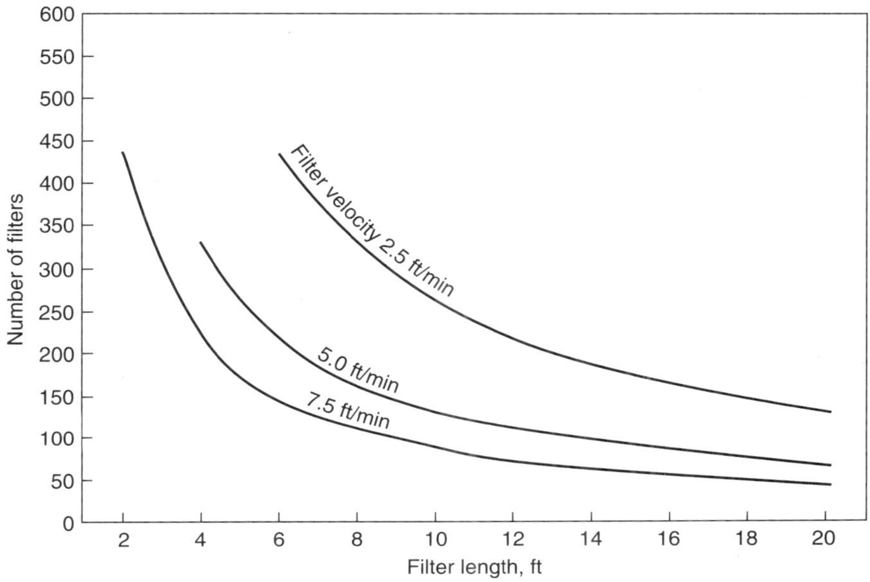

Figure 4.2 Optimizing number of filter bags versus filter bag length.

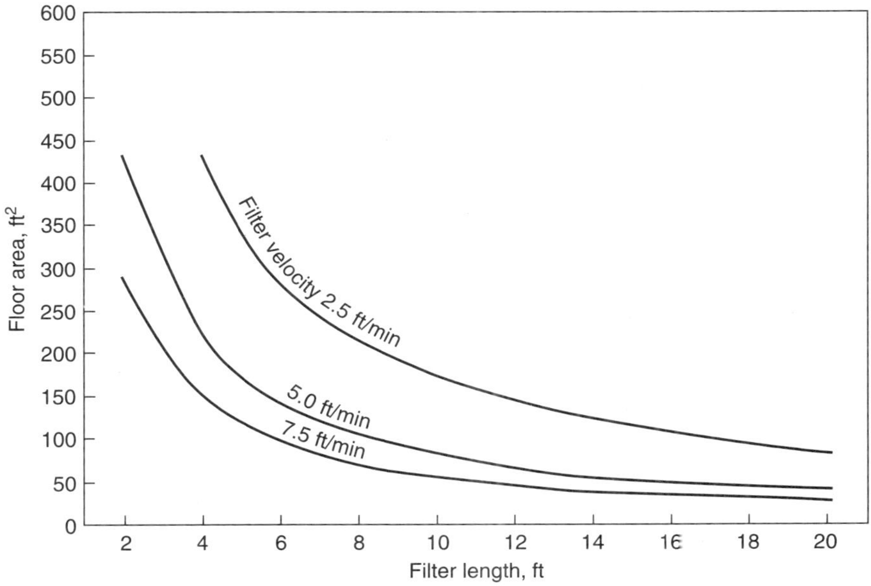

Figure 4.3 Optimizing floor area and footprint, versus filter bag length.

4.2 Optimum Filter System Design

Emissions regulations are changing rapidly, and technological advancements are being adapted to meet or exceed the ever-tighter pollution limits. Materials that once were utilized worldwide as the best available product technology, such as asbestos and freon, are now found to create world-class hazardous emissions and must be controlled with the best available control technology (BACT).

4.3 Optimum Collector Sizing from Figures

Notice in Table 4.1 that for the shorter 6- and 8-ft bags the 8-ft/min filtration velocity (A/C ratio) is limiting; while for the longer bags the limit is set by the maximum can velocity. In this case using the 6-ft bag gains nothing over the 8-ft one except a lower can velocity, and cost would be increased. Only if the lower can velocity is necessary to allow finer dust to settle out of the filter passage area should the 6-ft length be considered.

However, the step from 8- to 10-ft length is critical in this case since the total filter area increases at the limit of 250-ft/min can velocity. So the filtration velocity decreases, but the can velocity is 37 ft/min more. This tabulation from the charted figures (Figs. 4.1 through 4.3) may help to analyze the selection of an optimum size.

TABLE 4.1 Optimum Sizing Chart Data

Velocity*		Filter †			Footprint area, ft^2,‡ for the following filter spacings		
V_f, ft/min	V_c, ft/min	Length, ft	Area, ft^2	No.	10.5 × 9 in	9 × 9 in	10.5 in^2
8p§	160	6	1250	136	89	77	105
8p	213	8	1250	102	67	58	79
7.5	250p	10	1338	87	57	49	67
6.2	250p	12	1605	87	57	49	67
5.4	250p	14	1873	87	57	49	67
5.0	250p	16	2140	87	57	49	67
4.3	250p	18	2409	87	57	49	67
3.9	250p	20	2676	87	57	49	67

*See Fig. 4.1. Can velocity V_c will change if filter spacing changes [see Eq. (4.2)].

†See Fig. 4.2. (Filtration area)/(bag element) = 1.538 ft^2/ft for 6-in-diameter bags.

‡See Fig. 4.3. Footprint area = 0.656 ft^2/bag for 10.5- × 9-in spacing of bags center to center, 0.5625 ft^2/bag for 9- × 9-in spacing, and 0.7656 ft^2/bag for 10.5- × 10.5-in spacing. The (can passage area)/(filter bag) = (0.656 − 0.1964) ft^2/bag = 0.4596 ft^2/bag for 10.5- × 9-in spacing, 0.5625 − 0.1964 = 0.3661 ft^2/bag for 9- × 9-in spacing, and 0.7656 − 0.1964 = 0.5692 ft^2/bag for 10.5- × 10.5-in spacing.

§p indicates controlling parameter.

Nominal optimum size

For the maximum nominal filtration velocity of $V_f = 8.0$ ft/min, an 8-ft-long bag length requires a 67-ft² footprint, V_c is 213 ft/min which could be beneficial for reduced dust retainage on the filters, number of bags = 102, and cloth filtration area = 1250 ft².

However, the footprint could be reduced to 57 ft², by using 87 ten-foot-long bags, a can velocity V_c of 250 ft/min, a filtration velocity V_f of 7.5 ft/min, and 1338 ft² of filtering area. This same V_c could also use the 12-, 14-, 16-, 18-, or 20-ft-long bags. No purpose is served by a longer bag length or added filter area once the critical can and filtration velocity are met, unless there is a reason for a velocity reduction.

Thus the initially optimum size in this case would appear to be the 10-ft-long bags to utilize all the can velocity allowed of $V_c = 250$ ft/min and to reduce the footprint to 57 ft² and the V_f to 7.5 ft/min. The smaller footprint should lower housing cost, and the lower filtration velocity should reduce pressure drop across the bags.

Effective optimum size design variations

For the effective filtration velocity to be adjusted down if needed for operating factors, $V_{fe} = 4.6$ ft/min. With 12-ft-long bags and the $V_{fe} =$ 4.6 ft/min, from Fig. 4.1 the can velocity becomes 186 ft/min, from Fig. 4.2 the number of filter bags becomes 118, and from Fig. 4.3 the footprint area becomes 77 ft². The filter area = 118 × 1 538 ft²/ft × 12 ft = 118 × 18.456 ft²/bag = 2178 ft². Thus with volume $Q = 10{,}000$ ft³/min and $V_{fe} = 4.6$ ft/min, $Q/V_{fe} = (10{,}000$ ft³/min)/4.6 ft/min = 2178 ft², which checks the calculation.

Optimum size and design comparison. Table 4.2 summarizes four design and size combinations to approach an optimum for the 10,000 ft³/min airstream volume. Design 1 uses the minimum footprint area of 57 ft², minimum number of filter bags at 87, maximum can velocity

TABLE 4.2 Optimum Size and Design Comparison

Design parameters	Design 1	Design 2	Design 3	Design 4
Spacing	10.5 × 9 in	10.5 × 9 in	9 × 9 in	10.5 × 9 in
Number of bags	87	117	117	90
Bag length	10 ft	12 ft	12 ft	10 ft
Filter area	1338 ft²	2160 ft²	2160 ft²	1384 ft²
Volume Q	10,000 ft³/min	10,000 ft³/min	10,000 ft³/min	10,000 ft³
Footprint area	57 ft²	77 ft²	66 ft²	59 ft²
V_f	7.5 ft/min	4.63 ft/min	4.63 ft/min	7.23 ft/min
V_c	250 ft/min	186 ft/min	234 ft/min	242 ft/min

V_c of 250 ft/min, a slight reduction in filtration velocity V_f to 7.5 ft/min, and close to the minimum filter area of 1338 ft^2. However, the 87 bags is not a practical number for cleaning and blowpipe design, with an uneven number of valves and bags per row.

Practical pulse-jet design and cleaning configuration. However, design 2 for a practical design of the 118 cleaning valve and blowpipe configuration, uses 117 filters, which is divisible into 9 valves and blowpipes to each serve 13 bags, so 9 × 13 = 117 bags. Pulse-jet cleaning valves are available to handle up to 18 filters per valve and still maintain effective cleaning. Design 2 also achieves the 4.63 ft/min filtration velocity where that is necessary. Typical configurations will vary with

- Type and length of filter elements
- Operating pressure of the cleaning air
- Use of diffusers, venturis, nozzles, or cage-diffusers to optimize cleaning effectiveness down the full length of the filters
- Design standards

Effective total cloth area. The effective total cloth area = 117 bags × 18.456 ft^2/bag = 2160 ft^2. The effective filtration velocity V_{fe} thus becomes (10,000 ft^3/min)/(2160 ft^2) = 4.63 ft/min which is significantly lower than the calculated 8 ft/min objective. The can velocity is (10,000 ft^3/min)/(117 bags × 0.4596 ft^2/bag passage area) = 186 ft/min. The footprint area using the 10.5- × 9-in spacing = (10,000 ft^3/min)/(117 bags × 0.656 ft^2/bag) = 77 ft^2.

Filter spacing optimization. For further optimization, design 3 reduces the bag spacing to 9- × 9-in centers to better utilize the allowable can and filtration velocities while reducing footprint area. Filtration velocity stays at 4.63 ft/min. The can velocity V_c becomes (10,000 ft^3/min)/(117 bags × 0.3661 ft^2/bag) = 234 ft/min, which is certainly acceptable in this case where 250 ft/min was the maximum allowed from experience. Then the footprint becomes 117 × 0.5625 ft^2/bag = 66 ft^2, which provides a cost reduction in housing and structure size and weight, and in vertical flat plate area in calculating wind and seismic loads when compared with the 77 ft^2 with the 10.5- × 9-in bag space. This reduced floor area will permit the fitting of a collector into a smaller plant layout space, which can be critical sometimes.

Examining the options further produces perhaps the closest to optimum combination in size, performance, and probable cost, design 4, where filtration velocity can be near nominal. This uses the 10.5- × 9-in spacing, 90 bags in a 9 valve arrangement with 10 bags per valve, and 10-ft-long bags for a 1384 ft^2 filter area. Design 4 produces a calculated footprint area of 59 ft^2, a filtration velocity of 7.23 ft/min, and a can velocity of 242 ft/min. Thus design 4 is probably the most practical combination and truly the optimum size and cost, if the higher filtration velocity is acceptable (Table 4.2).

Collector height optimization. A further reduction in weight and cost can be made by the use of a shorter-height clean-air plenum. While the *walk-in* design uses the height of the filter elements plus 6 in of clearance for replacement of filters, the *top-door* design only uses about a 2-ft-high clean-air plenum (CAP) to reduce both height and weight. In some cases this top-door design is preferred for bag and cage access without entering the CAP fully, for quicker venting of hot or noxious air or gas than with the walk-in type.

The modular design in Fig. 3.12 uses a *lift-off* top or CAP to provide the ultimate top access and fast clearance of hot or noxious air. The pulse-jet cleaning system and blowpipes are fastened to the CAP wall panels so that everything above the tube sheet, where the bags and cages are mounted, is removed in one assembly. A further refinement is the addition of one large shelter building or penthouse over the combined modules so that maintenance and filter replacement on one module can take place out of the weather while the other modules remain on line. A monorail and hoist can be mounted under the shelter roof to make removal of a top panel assembly an easy and safe procedure.

Objectives. Filter collectors should be designed for optimum pollution control performance for each application and to achieve the best available control technology (BACT) wherever possible. Their design should also aim for optimum energy and cost savings, as well as economic justification to make the future economically bright, provide jobs that create value-added output, and help pay off the national debt.

4.4 Designing a Filter Collector System

With all the previous discussion behind us, let's put together a system for a typical dust collection problem with the necessary accessories, and take a stab at optimizing it!

Filter collector specifications

Product	Cement powder
Density	80 lb/ft^3

Particle size	10 to 50 μ (median—15 μ)
Application	Product collection
Inlet loading	80 to 100 gr/ft^3 = 8572 lb/h [see Eq. (6.1)]
Emissions	Limited to 0.02 gr/ft^3
Volume Q	10,000 ft^3/min airstream
Nominal maximum V_{fn}	8.0 ft/min
Effective V_{fe}	4.63 ft/min
Maximum can velocity V_c	250 ft/min (but to achieve the V_{fe}, V_c drops to 186 to 242 ft/min, per Table 4.2 designs, which is conservative compared to experience)
Measured ductwork pressure drop	6 in WG
Airstream temperature	230°F, 110°C
Materials of construction	A36 carbon steel
Coatings	One prime and one finish coat for corrosive exposure protection to mild HCl inside and a coastal weather exposure outside
Housing	12 gauge minimum, reinforced for ½ in maximum deflection
Tube sheet	10 gauge minimum, reinforced for ½ in maximum deflection
Filter makeup	16-oz singed polyester felt snap-band bags
Cage makeup	Galvanized HRS, hot rolled steel, 10 vertical wires × 8-in ring spacing, ⅛-in-diameter wire
Diffuser makeup	24 ga galvanized perforated HRS ⅝ of bag length, 40 to 50% open area
Maximum negative pressure	18 in WG
Hopper	Will have airlock discharge
Maintenance	Access walk-in clean-air plenum, ladder, cage, platform, safety hinge gate
Cleaning controls	Photohelic on-demand, black polyethylene tubing to valves, prewire and pretube; local panel with on-off pushbuttons; collector air supply; RAL; fan; auto-man-timed cleaning solenoid switch
10-min delay timer	For cleaning and RAL operation after fan and dust input are turned off

Electrical enclosures	NEMA #4—3 phase, 60 Hz, 480 V ac, controls 1 phase 115 V ac, 60 Hz
Clearance hopper discharge	4.5 ft above grade
Wind velocity	120 mi/h
Seismic zone	1
Extras	Backward inclined fan and drive with TEFC (totally enclosed fan cooled), NEMA 4 motor, grounding studs on, tube sheet, grounding wires in bags, pressure-relief latches on dirty-air plenum access door, secondary charcoal filters after clean-air plenum
Shipment	By truck within 3 months of purchase (provide fan mount and ship as completely assembled as possible)

Bag filter solution

Using our previous calculations and the initial specifications above: nominal V_{fn} = 8.0 ft/min changes to an effective V_{fe} = 4.63 ft/min; maximum V_c = 250 ft/min, but use V_c = 234 ft/min to hold V_{fe} = 4.63 ft/min with the use of 6-in diameter × 12-ft-long bags and 9- × 9-in bag spacing. The number of bags = 117, so filter area = 117 × 18.456 ft²/bag = 2160 ft².

Dimensions. With 9- × 9-in bag spacing selected for a 117-bag collector, width = 9 rows of valves × 9 in center to center = 81 plus 3 in = 84 in or 7.0 ft wide; and depth = 13 bags/row × 9 in center to center = 117 in + 3 in or 10 ft deep.

Footprint area. Footprint area = 7.0 ft × 10 ft deep = 70.0 ft² actual.

Height. Height will be the sum of CAP* = 12.5 ft and DAP = 12 ft and hopper = (10.0 ft² × tan 60° − 1.0 ft = 7.66 ft = 32.16 ft. Therefore, the height of the OAH collector without a support structure is 32.16 ft. With a 4.5-ft clearance under the hopper discharge, the total height is 36.66 ft. So the filter collector can be shipped by truck assembled either way since the length laying down will fit a 40- or 45-ft-long trailer, and the load height will be 10 ft − 0.5 ft for reinforcing ribs = 10.5 ft + 3.75 ft for low flatbed height = 14.25 ft OAH over the road, and width = 7.0 ft + 0.5 ft = 7.5 ft = 90 in OAW.

Estimated weight. 117 bags, 12 ft long, 7 × 10 ft collector, 36.66 ft high with 4.5 ft clearance above grade.

*CAP = clean-air plenum, DAP = dirty-air plenum, OAH = overall height, OAW = overall width.

Housing
(12-gauge) weight = weight of material (= 4.38 lb/ft^2)

$\times$ perimeter length (= 10.0 ft + 7.0 ft $\times$ 2 = 34.0 ft)

$\times$ height [= 12.5 ft + 12 ft + 7.66/(sin 60°/2) = 32.16 ft]

+ top (roof) area (= 7.0 ft $\times$ 10 ft = 70.0 ft^2)

= 4.38 lb/ft^2 $\times$ 1163.44 ft^2

= 5096 lb

T/S (10-gauge) weight = weight of material (= 5.63 lb/ft^2)

$\times$ 7.0 + 0.5 $\times$ 10.0 + 0.5

where 0.5 ft = flange width

= 5.63 lb/ft^2 $\times$ 78.75 ft^2

= 444 lb

Reinforced rib weight = weight of material (= 5.4 lb/ft)

$\times$ perimeter/2.8 ft spacing

$\times$ (height + 7.66 ft/sin 60° + top)

= 5.4 lb/ft $\times$ 34 ft/2.8 ft $\times$ (32.16 + 8.84

+ 10 ft/2.8 ft $\times$ 7.0) = 5.4 $\times$ [(13 $\times$ 41 ft) + 28]

= 3030 lb

T/S rib weight = 5.1 lb/ft $\times$ (10 ft/2.33 ft $\times$ 7.0 ft + 7.0 ft/2.33 ft

$\times$ 10 ft) = 5.1 (28 ft + 30) = 296 lb

Structural steel, platform, ladder and cage

= 4 columns $\times$ 16 lb/ft (5 WF) $\times$ (7.66 + 4.5 ft)

= 779 lb (5-in-wide flange columns)

Cross bracing angles = 4 $\times$ 2/side $\times$ 13.5 ft $\times$ 3.19 lb/ft

= 345 lb (2 $\times$ 2 $\times$ ¼ angles)

Base and top plates = 8 × 0.45 ft^2 (8 × 8 in) × 20.42 lb/ft^2

= 73 lb

Ladder = (12 ft + 12.16 ft + 3.0 ft above T/S) × 21 lb/ft

= 560 lb

Platform = 7.9 ft long × 3.5 ft wide × 25 lb/ft^2 typical

= 692 lb

Bags, cages, diffusers = approx. 2 lb/ft typical × 12 ft × 117 = 2808 lb

Total estimated collector weight = All above = 14,123 lb

Total estimated collector cost (at \$1.50/lb) = \$21,185

Estimated collector cost/(10.000 ft^3/min) = \$2.12/(ft^3/min)

Cartridge filter collector for 10,000 ft^3/min

Using the same performance specifications, design a cartridge filter solution. The first change would be to reduce the V_f from 4.63 down to (±)2.3 ft/min for the cartridge filter elements. This should be further adjusted for the dust load of 80 to 100 gr/ft^3, which is 3 to 4 times the usual maximum of 25 gr/ft^3 for cartridge filters. So the 2.3/2 = 1.15 ft/min should be reasonably conservative.

But based on a typical cartridge spacing, the V_{fe} could be 2.0 ft/min for a V_c of ± 243 ft/min. Suppliers should verify the performance that they will guarantee, especially on fines that may cake excessively in the pleats. So based on grain loading and pleat caking, we will use the average of (1.15 + 2.0)/2 = 1.575 ft/min.

At 1.58 ft/min then, the filter area should be (10,000 ft^3/min)/(1.58 ft/min) = 6330 ft^2. Using standard collector sizes from one manufacturer with 300 ft^2 per cartridge, we can provide 22 cartridges with 6600 ft^2. This provides a V_{fe} = 1.515 ft/min and a V_c = ± 184 ft/min. The next smaller is 6000 ft^2 which produces a V_{fe} of 1.67 ft/min and a V_c of 203 ft/min. This might well be acceptable but not without verification from the supplier or from tests, due to the potential caking in the cartridge pleats, and the inlet loading at 3 to 4 times normal for cartridges. So stick to the 6600-ft^3/min size and get bids based on supplier guarantees for this performance and for acceptable alternatives.

Alternatives would include evaluating other filter cartridge designs and performance. The industry standard has previously been 240 to

300 ft^2 per cartridge [at (12.75-in diameter × 3.1416)/(300 pleats) = 0.1335 in maximum spacing of pleats at the outer edge.] This sets the space between the pleats at about ⅛ in maximum at the circumference after deducting the filter media thickness. Thus the tendency for dust to cake between the pleats is obvious, especially closer to the central core where spacing is less.

One of the alternatives would be to use wider spaced folds in the cartridges, and these are now available down to 50 pleats per cartridge [again at (12.75-in diameter × 3.1416)/(50 pleats) = 40.06 in/50 = 0.801-in maximum spacing between pleats]. This ¾-in nominal spacing after deducting the filter media thickness should allow the filtration velocity to be raised, possibly up to around 4.2 ft/min. Using that V_{fe} produces a filter area required of (10,000 ft^3/min)/(4.2 ft/min) = 2380 ft^2. However, these cartridges have only 36 ft^2 each of filtration area, so (10,000 ft^3/min)/(36 ft^2) = 278 cartridges. With an area of 2.16 ft^2 of footprint apiece, this will require 599 ft^2 of floor area, which is hardly practical at something like 52 ft wide × 11.5 ft deep.

Footprint and dimensions. The standard cartridge collector with 22 cartridges, each with an area of 300 ft^2, has a total area of 6600 ft^2. Without the fan, the collector's dimensions are 180 in (15.0 ft) wide × 53 in (4.42 ft) deep, and 126 in (10.5 ft) high. The floor area or footprint is 15.0 ft × 4.42 ft = 66.3 ft^2. If height or weight is a factor, then the cartridge has a distinct advantage, as detailed in the comparison shown in Table 4.3 (see Fig. 4.4).

TABLE 4.3 Dimensional Comparison of Collector Design Solutions

Design solution	Area, ft^2	Height, ft	Weight, lb
Bag collector	10.0 × 7.0 = 70.0	32.16	14,123
Cartridge collector	15.0 × 4.42 = 66.3	10.84	5,400
Modular collector	9.0 × 9.0 = 81.0	24.00	12,000

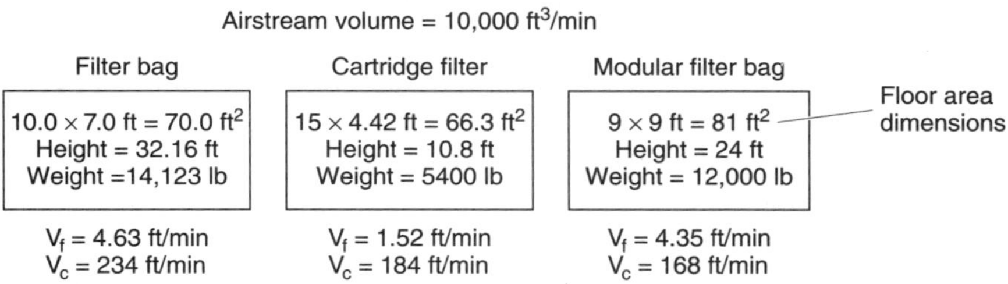

Figure 4.4 Comparison of optimum filter bag collector, optimum cartridge filter collector, and optimum modular filter bag collector.

Building block modular solution

Using the V_{fe} of 4.63 ft/min, we found (10,000 ft^3/min)/4.63 = 2160 ft^2 filter area required. The closest standard module is 2298 ft^2. This is a 110-bag collector with 14-ft-long bags. Dimensions are 9 × 9 ft with a 24-ft height, using a short 2-ft-high clean-air plenum. The other performance figures should be fairly similar, or more conservative in this case with more filter area than the bag filter which is essentially custom designed (see Fig. 3.12).

Changes in size would be limited for each module without sacrificing the cost savings of standardized design and fabrication. However, in collectors with larger airstream volumes, the modular construction should be a great advantage and meet the optimum sizing much more readily.

4.5 Collector Accessories

The necessary, and/or alternative, accessory equipment will be covered in subsequent chapters.

Chapter

5

Secondary Filtration

Secondary filtration accomplishes a finer step of collection than bag or cartridge filter media usually do. Those primary filter media will collect dust and other particulates down to the 0.5-μ particle size, at efficiencies of 99.99 percent and better depending on the filter cake. High-efficiency particulate air (HEPA) filters will capture particles, fumes, smoke, and gaseous molecules as small as 0.3 μ at efficiencies around 99.99 percent. Ultralow penetration air (ULPA) filters can extend the capture range of particles down to the 0.12-μ size at efficiencies up to 99.9995 percent. All filters may capture contaminants in finer sizes than listed with an appropriate dust cake on the filter surface. Secondary filters perform best when exposed to a very light dust load. HEPA and ULPA filters generally have no on-line cleaning provided.

5.1 Filtration after Primary Collection

Secondary filtration is a growing requirement as dust collection systems face ever tighter emission limits on exhaust to the atmosphere. Makeup air return to the inside of a plant or for process use and clean room air supply have also increased the need for secondary filters with better collection of fines. Secondary filtration is also increasing in industry to achieve a finer size separation and with provision for on-line cleaning and continuous performance. This secondary filtration is accomplished with bag and cartridge filter collectors that follow precipitators, scrubbers, and primary filters.

TABLE 5.1 HEPA and ULPA Efficiency

Efficiency, %	Particle size, μ	Face velocity, ft/min
HEPA 99.97 to 99.99	0.3	275
HEPA 99.97	0.3	500
ULPA 99.999	0.3	275
ULPA 99.9995	0.12	100

HEPAs

High-efficiency particulate air filters are the most common secondary filters. These must capture a minimum of 99.97 percent of the airborne particulates down to 0.3-μ size and are available for collecting up to 99.99 percent. Their face velocity rating is measured at a pressure drop of 0.5 in WG, but are operated at velocities up to 275 ft/min with initial clean resistance of 1.0 in WG. Higher velocity rating HEPAs are available at up to 500 ft/min with an initial clean resistance of 1.4 in WG.*

ULPAs

Ultralow penetration air filters are available that provide a minimum efficiency of 99.999 percent (0.001 percent penetration) on 0.3-μ particles. The ULPA filters operate at velocities up to 275 ft/min with an initial resistance of 1.0 in WG. Some ULPA filters are available at efficiencies up to 99.9995 percent on particles as small as 0.12 μ. (See Table 5.2.)

*Barneby and Sutcliffe Corp., Columbus, OH; Bonaire Filtration Technology, Inc., Greensboro, NC 27419; Flanders Filters, Inc., Washington, N.C. 27889.

TABLE 5.2 HEPA and ULPA Standard Sizes

Standard sizes* (H × W × D)	Air flow, (ft^3/min)/(ft/min)	Standard initial pressure drop, in WG	Particle size, μ	Particle size efficiency, %
12 × 12 × 6	150/150	1.0	0.03	99.99
24 × 12 × 6	300/150	1.0	0.03	99.99
24 × 24 × 6	600/150	1.0	0.03	99.99
24 × 24 × 12	1100/275	1.0	0.03	99.99
24 × 30 × 12	1450/290	1.0	0.03	99.99
24 × 12 × 12	1000/500	1.4	0.03	99.97
24 × 24 × 12	2000/500	1.4	0.03	99.97
24 × 24 × 5.875†	400/100	0.55	0.12	99.9995

*These are some typical standard sizes, and many other standard and custom sizes are available. (H × W × D) = (height × width × depth).

†Special low pressure.

DOP tests

HEPA secondary filters are required by industry-accepted standards to capture a minimum of 99.97 percent or up to 99.99 percent, or for the ULPA an efficiency of 99.999 percent, of thermally generated monodispersed 0.3-µ particles. This means that they should permit no more than 0.001 to 0.03 percent of the particles to penetrate the filter. This is a U.S. Government–approved test method using di-oxy-phthalate (DOP) for filter performance. This test is conducted in accordance with the *Edgewood Arsenal Instruction Manual #136-300-175* using a Q.107 penetrometer (forward light scattering photometer). Filters can also be tested using ambient air or other test methods where DOP testing is not permissible or applicable.

Leak tests

Filters should also be leak-tested by the manufacturer to ensure that they have no pinhole leaks (IES-RP-CC-001-86). Subsequent field testing is available from filter suppliers to certify that filter systems continue to meet performance specifications in operation. Gaskets and sealants are used to ensure all-around leakproof service in operating conditions.

Temperature and humidity

Materials used in standard HEPA and ULPA filters are rated for continuous service at temperatures up to 250°F (121°C), and at a relative humidity of up to 100 percent. High-temperature materials are available for filter service up to 500°F.

Tertiary-stage carbon filters

Tertiary-stage carbon filters, usually of the carbon or carbon-alumina type, are required in some systems to adsorb gaseous fumes, odors, and very fine particulates not captured by the secondary stage. The American Society of Heating, Refrigerating, and Air Conditioning Engineers code 62-89 requires a minimum of 20 ft^3/min per person of fresh makeup air in a typical office environment, or 60 ft^3/min in smoking areas. Odor control is largely a function of the activated carbon density in the filter, or carbon-alumina combination. This density typically ranges from 100 to 375 g/ft^2 of filter face area.

Each granule of activated carbon contains a maze of microscopic pores slightly larger than the size of the gas molecules adsorbed. This porous structure produces a very large internal surface area that captures the gas molecules. This internal area should run from 1,320,000 to 3,960,000 ft^2 per square foot of carbon filter as the density is

TABLE 5.3 Carbon Filter Performance

Initial resistance, in WG	Face velocity, ft/min	Average arrestance @ 500 ft/min, %	Recommended final resistance, in WG	Carbon density, g/ft^2
0.17	300	60–65	1.0	55/100
0.30	500	60–65	1.0	55/100
0.19	300	70–75	1.2	200
0.40	500	70–75	1.2	200
0.28	300	75–80	1.2	300
0.68	500	75–80	1.2	300
0.40	300	80–85	1.2	600
0.75	500	80–85	1.2	600

increased. Operating temperature limits of carbon filters are usually up to 120°F (49°C).

Carbon panels are produced in 1- and 2-in thickness and in the same dimensions as HEPA filters to fit into the same cross-sectional size housing racks (see Table 5.3).

5.2 Alternative Filter and Collection Means

Other collectors for capturing or purifying fumes, odors, mists, sprays, and fine particulates will need more thorough description than presented here, but they are outlined briefly.

Alternative secondary filters

Other types of secondary filters are available using synthetic fibers or other media, but often with lower collection efficiencies. A micro-fiberglass paper media is serviceable up to 350°F (177°C), at an efficiency of 95 percent with 0.3-μ particles and with relative humidity to 100 percent. These and others under development may meet the actual filtration need and even provide longer life than HEPAs.

A face velocity of around 250 ft/min in a 6-in-thick panel produces an initial pressure drop of 0.90 in WG and a final recommended resistance of 1.5 in WG. A face velocity of 500 ft/min is available in a 12-in-thick panel which produces an initial pressure drop of 0.90 in WG and a final recommended resistance of 2.0 in WG. Standard panel sizes match the HEPA and carbon filter sizes.

- *Electrostatic, ion, and ozone collectors.* For smaller air volumes the ion collector attracts particles and the ozone generator oxidizes organic gas molecules, thus reducing odors, gases, and volatile organic compounds (VOCs) to harmless CO_2 and O_2 as found in clean air. Electrostatic precipitators are used on larger-volume airstreams for collecting fly ash and similar dusts, usually as primary collectors, often followed by higher-efficiency filters.

- *Wet scrubbers.* Wet scrubbers are used especially following wet processes like spray driers and for cooling incinerator and process airstreams. But filter collectors are finding applications here with their advantage of greater efficiency and the availability of higher-temperature filter media that eliminates the need for heat exchangers or dilution air, as well as the need for proper disposal or recycling of wastewater from scrubbers.
- *Water-wash spray booths and/or washable filters.* These are used for primary filtration of wet overspray and mists.
- *Neutralizing additives or sorbents.* These additives, such as ground limestone, reduce acid concentration in the airstream, while also serving as a filter aid or precoat on the filter surface. This controlled filter cake provides a neutralizing action during the airstream residence time on the filter to both reduce acid concentrations in the flue gas and to improve filtration. (See Chap. 7, Lime Injection in Accessories.)

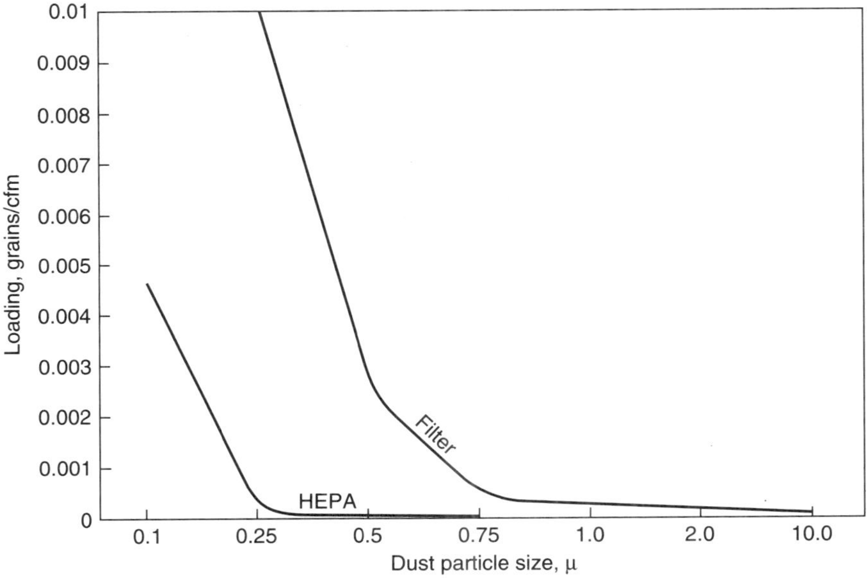

Figure 5.1 Dust emissions loading for primary and secondary collection.

5.3 Application of Secondary Filters and Alternatives

Figure 5.1 shows typical dust emissions loading for collectors in series for primary and secondary collection. Most industrial secondary filters are used after a high-efficiency filter or other dust collector, so they are only exposed to particles smaller than 0.5 μ and to a low dust loading. When used after a low-efficiency collector, or with no primary collector as in a heating and ventilating application, a row of prefilters should be added ahead of the secondary collector. The prefilter is usually a 1- to 2-in-thick panel in standard sizes or roll-type filter which collects primarily the larger particles and thus minimizes the dust loading on the secondary filters. (See Table 5.2.)

For air recirculation inside a building, it is likely that the Occupational Safety and Health Administration (OSHA) will require the use of these high-efficiency or HEPA filters, or the same results, per the smoke inhalation guidelines above. For liability protection it makes good sense to use an effective HEPA-type secondary filter to protect any employees that may be exposed to inhalation of fine particulates, even if not currently required to make provision for them.

Filter safety monitor

By pressure-drop measurement across the secondary filters a beneficial auxiliary use is available. A photohelic gauge controller can be installed on a bank of secondary filters to indicate the differential pressure measured across the HEPAs and provide a control switch action. Thus if the normal operating pressure drop is 1 to 2 in WG, the control point might be set from 2 to 3 in WG. When that set point is exceeded, an alarm or control action or both can be initiated. This provides an automatic warning of overloaded secondary filters due to a break in the primary filters or overloading from long-term normal exposure of the secondary filters.

Housings for secondary, tertiary, and prefilters

Most manufacturers of secondary filters offer housings for multiple filter panels and for several rows of filters in sequence in the airstream. If the secondary filter selected is a 24 × 24 × 12-in panel that has a flow capacity of 2000 ft^3/min and the airstream volume is 10,000 ft^3/min, the minimum number of panels required is five. Most housings are made in sizes that are multiples of two. So a good choice would be a housing of six panels mounted two rows high and three panels across each row. This provides 2000 ft^3/min of extra capacity, which should pay for itself in a longer filter service life and reduced pressure drop across the filter bank (see Fig. 5.2).

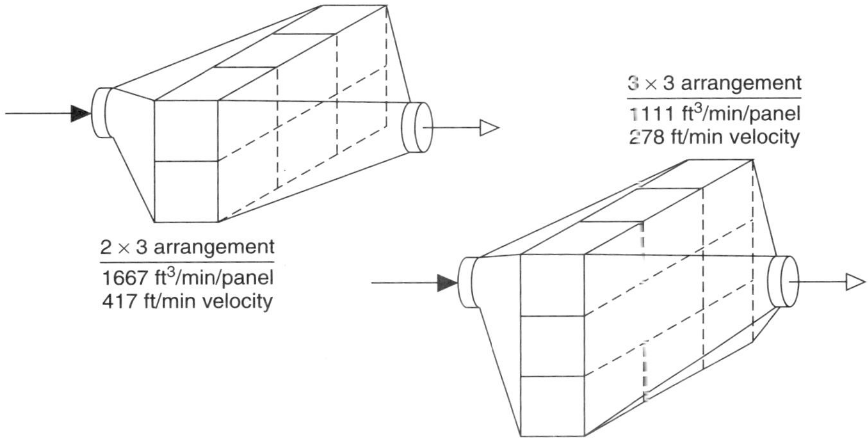

Figure 5.2 Secondary filter housing design.

Filter requirements will determine the number, type, and sequence of filters. The primary filter or prefilter would be a filter collector for most process or general industrial collection. For most heating, ventilation, and air-conditioning purposes, a panel-type prefilter 1 or 2 in thick would be most likely. For vapor and fume control, a tertiary row of carbon filter panels is usually added (see Table 5.4).

The housings are usually fabricated of sheet metal and designed for mounting in the ductwork after the primary collector. Carbon, galvanized, and stainless steel are the most common materials of construction along with aluminum or particle board. The housings are typically made in a thickness of 12 or 14 gauge, unless the process pressure demands heavier construction. Gaskets are used to seal the access

TABLE 5.4 Typical Filter Combinations

Primary, secondary, tertiary	Application
Prefilter, HEPA	Particulate filtration
Prefilter, HEPA, carbon	Particulate filtration and vapor
Prefilter, carbon	Vapor only
Prefilter, HEPA, carbon, HEPA	Intake particulate and vapor into a controlled space; for toxic gases use two carbon filters
Prefilter, carbon, HEPA	Intake and recirculation vapor only

Note: Check insurance, OSHA, and safety requirements as well as pollution regulations. Again, "On the average, nothing is average."

doors for the pressure involved and are selected for the service temperatures and airstream exposure conditions.

The filter housings are available with slide-in holders or racks for prefilters and secondary and tertiary filter panels. Pipe fittings for pressure-drop measurement can be built in. Structural supports for flange mounting are usually supplied in the field by the ductwork contractor. In most cases adequate transitions should be provided to expand from the ductwork size to the larger secondary filter cross-sectional area and to reduce the area back to the duct size, with minimum entry losses. (See Chap. 8 and Fig. 8.3 on ductwork, hoods and transitions, and air movement.)

The normal configuration has the filters standing up vertically perpendicular to the airflow. When this arrangement becomes too large in cross-sectional area, the filter panels can be arranged in a V pattern. The panels will be placed at 15° to 30° above horizontal in slide-in racks that permit installation of about twice as many filter panels in the same vertical space. An alternative arrangement would be placement at 15° to 30° angles from the airstream centerline, but with this arrangement access may not be as easy as the horizontal V layout. Either the HEPA suppliers or ductwork contractors can quote on the specific housing needs and layout arrangements.

Modular and standardized housing designs are available in building block configurations. A 1×1 arrangement will measure about $24\frac{1}{8}$ in (width) $\times$ 30 in (height) $\times$ 24 in (thickness). A 12-filter arrangement in a 4×3 pattern will measure about $72\frac{1}{8}$ in (width) $\times$ 120 in (height) $\times$ 24 in (thickness). These are for just one row deep of HEPA and prefilter panels. The addition of carbon filters and multiples of any filters in sequence will add thickness along the airstream. Dimensional variations for different manufacturers will occur, so each must be verified.

Local or self-contained collectors

Self-contained, portable, or mobile air cleaners for hazardous pollutants and dust are available for localized and work station collection where a central system is not economically feasible. These local collectors are used for such service as vacuuming up lead dust with a built-in HEPA filter for 99.99 percent efficiency for 0.12-μ particle sizes. Other uses would be for temporary or unusual local needs for dusts from sanding fiberglass, sawdust, and welding smoke and fumes. *Note:* Any hazardous collection must meet EPA and OSHA regulations, and thus cleaners must be designed to meet these requirements. Clean rooms and carcinogenic and other specialized collection should be carefully checked out. Localized hoods for spray, laboratory work, and other vapor or fume collection are a consideration where they too meet applicable regulations.

Chapter

6

Feeders and Discharge Equipment

Dust and bulk solid materials must be introduced into and discharged from dust collectors at a uniform rate while maintaining a pressure or vacuum seal. Various types and designs are used successfully in different applications for feeding and discharging, but describing them all as feeders will suffice except where special discharge requirements are encountered. Feeders should be selected on the basis of past experience or tests with the material to be handled to determine suitability, accuracy, and cost-effectiveness.

Feeders can be run fully loaded so should be designed for 100 percent loading where there is no other feed rate control. Most processes will have upstream infeed rate controls, so the dust collector will normally be fed at a uniform rate. But watch for situations where dust is accumulated in the collector hopper at shutdown and a 100 percent volume load must be handled at start-up. In that case the drive, shaft, and all mechanical parts must be sized for full loads, or a slow speed or low torque start-up controller used. Special motor controls may be used to ramp the current or torque so there is a slow start-up and normal horsepower components used, but consider the drive section below.

There are two main types of feeders:

Gravimetric feeders. A feeder that weighs the material fed or discharged can usually control feed rate to an accuracy of ½ percent or less and is called a gravimetric feeder. Normally these are used only on material formulation or other processes where accuracy is critical. The cost of a gravimetric feeder is around 2 to 3 times more than a volumetric feeder.

Volumetric feeders. Most dust collector systems control feed rate by the volume moved or conveyed and thus use volumetric feeders which can control delivery rate to an accuracy of around 1 to 3 percent. Sometimes there is a need for an accuracy around 1 percent, and some volumetric feeder designs can fill the bill. However, the incoming feed supply actually controls the rate possible to some extent, so the discharge feeder can only meter dust within the supplied volume range and seal the pressure. Specific types of feeders are discussed in the following sections.

6.1 Rotary Air Locks

This is one of the most commonly used feeders. Its relatively low cost, dependability, simplicity, and generally adequate leakage seal make it the standard feeder. But beware of leakage! In the best of conditions almost all rotary air locks (RALs) will bypass or leak some 5 to 15 ft^3/min depending on operating pressure and internal clearances. Rotary air locks are used on pressure differentials up to 10 lb/in^2 gauge and sometimes more with careful design and construction. (See Fig. 6.1.)

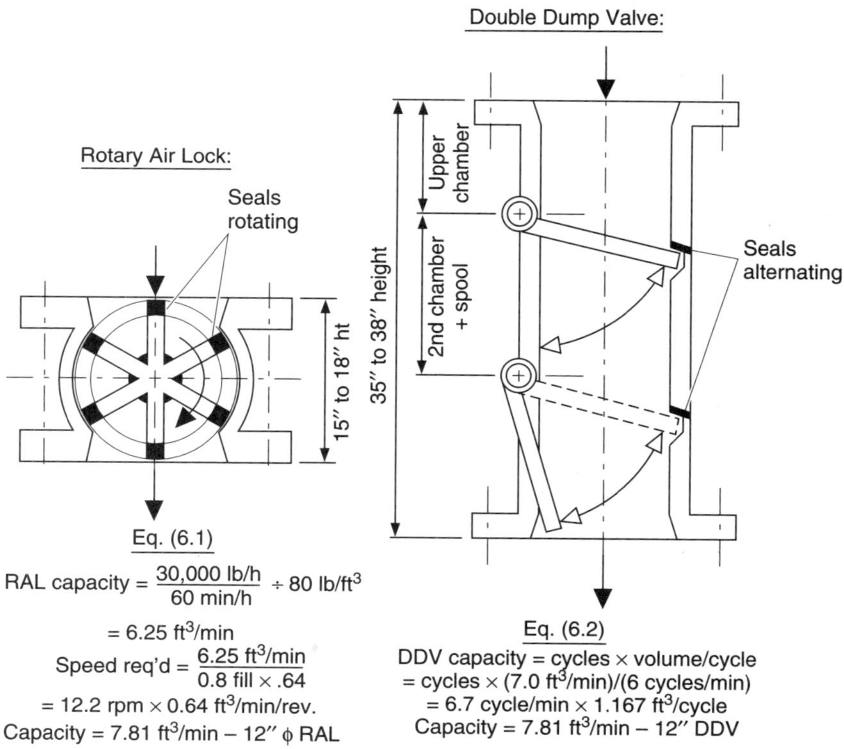

Figure 6.1 Feeder types: rotary air lock and double dump valve.

Abrasive materials and pressure differentials above 1 or 2 lb/in^2 gauge can often combine to cause leakage and abrasive wear in a RAL very quickly. Once leakage and wear start, especially with more abrasive dusts, wear can accelerate rapidly. Past experience and guaranteed recommendations from suppliers should be used as a guide for selection.

Design options to consider on RALs are air-purged shaft seals to prevent dust from entering the shaft and seal clearance openings and causing wear; closed-end rotors to minimize air bypass and rapid abrasive wear of the rotor and housing ends; adjustable tips or seal strips on the rotor vanes to reduce air leakage and wear as well as help clean dust off the rotor housing; hard cast-iron or chrome lining on the housing interior to reduce wear; and slower speeds to reduce the rate of wear. High-temperature operation requires suitable seals and outboard bearings to allow the RAL to withstand heat usually above 250°F (121°C). Clearance allowances for expansion and contraction will permit RAL operation up to 750°F (399°C). Corrosive products may require stainless-steel or Inconel construction or a chrome lining, as well as suitable seal materials.

RAL sizing

While RAL sizes are available in diameters from 4 to 30 in (101 to 762 mm), most collectors used are 8, 10, 12, or 14 in. Electric motors usually drive RALs, and power will run from ½ to 3 horsepower (hp) in these sizes. Speed of operation is best limited to a 15 to 20 rpm maximum, both to reduce abrasive wear and to provide some capacity available for future growth. The maximum normal speed is usually 25 rpm. Some air locks for dust collector service are run at up to 38 rpm with correspondingly increased capacities, such as on sawdust. But check whether the experience on wear is excessive with the dust or material to be run and whether it can be guaranteed. (See Table 6.1.)

TABLE 6.1 RAL Size and Capacity

RAL diameter, in	Capacity, ft^3/min						Drive horsepower at 15 rpm (estimated)
	ft^3/r	5 rpm	10 rpm	15 rpm	20 rpm	25 rpm	
4	0.02	0.1	0.2	0.3	0.4	0.5	⅓
6	0.07	0.35	0.7	1.05	1.4	1.75	⅓, ½
8	0.18	0.9	1.8	2.7	3.6	4.5	½–1
10	0.36	1.8	3.6	5.4	7.2	9.0	½–2
12	0.64	3.2	6.4	9.6	12.8	16.0	¾–2
14	1.12	5.6	11.2	16.8	22.4	28.0	1–3
16	1.62	8.1	16.2	24.3	32.4	40.5	1–3
18	2.29	11.45	22.9	34.35	45.8	57.25	1½–5
22	4.34	21.7	43.4	65.1	86.8	—	1½–5

Each successive feeder or conveyor in a handling system should be sized about 15 percent larger than the upstream unit to allow for adequate handling of surges and to prevent choking the flow. Some cases may need 20 to 30 percent greater capacity, especially if large surges in flow rate are possible upstream. To allow for partial filling of each RAL pocket versus the theoretical volume, allow more capacity than the theoretical requirement with a 75 to 80 percent fill factor (FF).

$$\text{Nominal RAL capacity} = \text{rpm} \times \text{cubic feet per revolution/fill factor} \tag{6.1}$$

For a typical case, say portland cement at 80 lb/ft^3 density and 15 tons per hour (TPH) required, capacity would be as follows:

$$\text{RAL capacity required} = \frac{15\ \text{TPH} \times 2000\ \text{lb/ton}}{80\ \text{lb/ft}^3}$$

$$= \frac{30{,}000\ \text{lb/h}}{80\ \text{lb/ft}^3}$$

$$= 375\ \text{ft}^3/\text{h}$$

$$= 6.25\ \text{ft}^3/\text{min required}$$

Then, from Eq. (6.1)

$$\text{Nominal RAL capacity} = 6.25\ \text{ft}^3/\text{min}/0.8\ \text{FF} = 7.82\ \text{ft}^3/\text{min}$$

We must allow for less than complete filling of the rotor pockets and some variation in input flow rate by the use of a fill factor (FF). An 80 percent FF is a good allowance from experience.

The recommended solution would be a 12-in RAL operating at 12.2 to 12.5 rpm, according to Eq. (6.1).

$$\text{Speed} = \frac{7.82\ \text{ft}^3/\text{min}}{0.64\ \text{ft}^3/\text{rev}} = 12.22\ \text{rpm}$$

As mentioned before, an initial operating speed of 15 rpm or less is recommended for reduced wear and to allow some capacity margin for future increase. On less-abrasive products and intermittent or short-life projects, the speed could be raised to 20 to 30 rpm. The choice between initial investment and maintenance cost on RALs or most other collection equipment is a case of "Pay me now, or pay me later."

Cement, being an abrasive dust product, should have some wear protection beside the restrained operating speed, such as a closed-end rotor, air-purged shaft seals, hardened adjustable seal strips or rotor vane tips, high-temperature seals and outboard bearings as the tem-

TABLE 6.2 RAL Budget Costs

Size, in	Cast iron	Abrasive seal	Closed end	Drive, 15 rpm	Motor, 15 rpm	500°F kit	Air purge	Total cost, $
4	535	116	116	410	114	22	67	1380
6	678	129	120	424	120	27	72	1570
8	833	143	148	472	143	36	76	1851
10	1078	152	179	557	148	46	81	2168
12	1413	170	192	678	161	59	94	2767
14	1965	179	224	847	174	72	103	3564
16	2456	201	250	909	174	85	116	4191
18	3239	224	341	1213	255	98	135	5505
22	3997	246	429	1386	255	148	148	6609

Note: These are 1993 estimated costs! Drive includes power transmission chain and guard, motor is TEFC (totally enclosed fan cooled enclosure), and total cost is an estimated budget figure. Other options will make a cost difference.

perature requires. Actual plant experience with maintenance and specific manufacturer's products should influence the selection. If this is a totally new application, get several manufacturers' recommendations.

The estimated budget cost for the 12-in RAL will be $2,767.00 if we take all the options listed for the cement application, which is recommended. (See Table 6.2.)

Optional RAL designs

There are many options available for adaptation to RAL design. One of these is the Rotoscoop for its self-cleaning design. The rotor has concave or cusped pockets which are cleaned on every revolution by a rotating cleaning blade. The rotor and blade rotation is synchronized by gears, so the cleaning blade wipes difficult or damp material out of each rotor pocket as it passes over the discharge opening.

Another is the shallow pocket rotor for improved rotor discharge and reduced volume per revolution, where very low capacities are needed, such as for lime injection into the dust collector inlet airstream. A 4-in RAL rotor with 24 shallow cut pockets, similar to a radial saw blade cross section, has been used to reduce capacity to 0.00058 ft^3/rev. It used a $\frac{1}{4}$ hp direct current (dc) variable speed drive and controller in a NEMA 12 enclosure to deliver a 20 to 1 speed range or turndown ratio with a 2 to 40 lb/h capacity on 60 lb/ft^3 lime. Cost increased from a budget of $1380 to $2200.

Air locks are nearly sparkproof in their standard construction because of the low rotating speed. When compared to a fan with a tip speed of around 10,000 ft/min, a 12-in air lock rotating at 15 rpm would have a tip speed of about 47 ft/min, or 0.8 ft/s. While this should be well below the speed to strike a spark, other possibilities exist. If static charge buildup between the rotor and housing is possible, then a flexible grounding connection, strap, or even a steel-wool pad can be used to conduct the static charge to a safe ground point. If a falling bolt, wrench, or piece of metal could fall on the RAL housing or rotor during operation, it might well cause a spark. A deflector or an all-brass construction might be used, but the odds on sparks happening should be discussed with the user's insurance carrier and the RAL manufacturer as to whether any protection beyond standard construction is necessary.

6.2 Double Dump, Tipping, or Flapper Valves

These are all names for a type of valve designed to use gravity flow through two vertically located chambers that operate alternately in sequence so that when one is open, the other is closed. This maintains an air seal at all times, with the alternately closed dump plates sealing up against a ground metal surface or a flexible gasket seal on the chamber above. The dump plates are usually cycled up to 6 times per minute so that capacity is equal to 6 times the volume of the lower chamber and spool piece.

Actuation is typically initiated by a pneumatic cylinder or motor drive linkage that opens one dump plate while the other is closed. Some are operated up to 10 cycles/min. Gravity operation using counterweighted arms and linkage of the dump plates is possible. The upper dump plate is opened when the weight of dust material exceeds the counterweight force. The counterweight force closes the upper dump plate when the dust is discharged. The bottom plate opens after the upper plate closes and discharges the batch of dust material.

The double dump valve is finding application on abrasive and high-temperature materials, such as fly ash, cement clinker, and stringy materials with the following advantages:

- Flat contact seating surfaces for a tight seal
- Superior wear resistance on abrasive materials
- High-temperature service capability
- Easy access for maintenance
- Handles pressure differentials up to 30 lb/in^2 gauge
- Reduced air leakage, approximately equal to the lower chamber volume times the cycle rate

The main disadvantages are

- Overall height is about 2 to 3 times the height of a RAL. For instance a 10-in RAL is about 15 in high and a 10-in double dump valve is about 40 to 44 in high.
- Investment cost is also about 2 to 3 times as much as a RAL.
- It cannot be used under a large head or weight of material.
- It cannot be used for metering as smoothly as a RAL or screw conveyor since it delivers a larger slug of material on each cycle.

Capacity

A 1-ft^3 upper chamber double dump valve operating at 6 cycles/min would deliver 6 ft^3/min. Larger capacity requires either a bigger chamber or a faster cycle. Another way to increase capacity is by inserting a spool piece between the two chambers. A 6-, 12-, or 18-in height is normally used and increases the holding volume of the lower chamber. The cycle would be timed to hold the upper plate open until the lower chamber and spool piece are nearly full, thus increasing the volume per cycle. One caution for this valve and most bulk solid material handling is that the dust must be free flowing enough to flow as calculated!

The 7.82 ft^3/min determined above for a RAL handling 15 tons/h of clinker cement included an allowance for a practical fill factor of 20 percent resulting in a theoretical operating capacity of 125 percent or 18.75 tons/h. To calculate a double dump valve (DDV) capacity and size for the same rate:

$$\text{DDV capacity} = \text{standard capacity} + \text{spool volume} \qquad \text{(at 6 cycles/min)}$$

$$\text{12-in basis capacity} = 10.0\ \text{ft}^3\text{/min} \qquad \text{(See Table 6.3)} \tag{6.2}$$

So we could use a 12-in double dump valve with a 12-in-high spool for a capacity of 10.0 ft^3/min, if the height of 36 in standard plus 12-in spool height = 48 in OAH is acceptable. This 12-in valve would deliver a weight = 10.0 $ft^3/min \times 80\ lb/ft^3$ = 800 lb/min, or 48,000 lb/h, or 24 tons/h.

Since this capacity is more than adequate, the cycle rate might be reduced from 6 cycles/min maximum. From Eq. (6.2), for DDV at 18.75 tons/h required from the RAL sizing:

$$\text{Rated capacity} = \frac{\text{24 tons/h}}{\text{6 cycles/min}} = \frac{\text{4 tons/h}}{\text{cycles/min}}$$

TABLE 6.3 Double Dump Valve Capacities, Height, and Cost

Inlet size, in	Capacity, ft^3/min			Standard height, in	Budget cost, $
	Standard	6-in-high spool	12-in-high spool		
6	0.6	0.9	1.2	27	2800
8	1.5	2.5	3.5	30	3175
10	3.0	4.8	6.7	33	3775
12	5.0	7.0	10.0	36	4140
16	11.0	15.0	21.0	42	5700
20	27.0	31.0	35.0	60	7100
24	50.0	58.0	66.0	84	8300

Note: Capacities are based on a maximum of 6 cycles/min and a maximum fill of material. A fill factor of 80 to 85 percent should be used for actual operating rates, as flow into the discharge valve can be affected by aeration, hopper flow and angle of repose, and uniformity of input flow. Spool costs can be estimated at approximately 10 percent of the standard valve budget cost for the 6-in-high and at 15 percent for the 12-in-high spool.

Options are available for temperatures to 1900°F (1038°C), vacuums to 27-in Hg (13.2 lb/in^2 or 358 in WG); pressures to 300 lb/in^2 gauge.

$$\text{Required speed} = \frac{18.75 \text{ tons/h}}{4 \text{ (tons/h)/(cycles/min)}} = 4.69 \text{ cycles/min}$$

Thus the DDV can be operated at 4.69 cycles/min and deliver 18.75 tons/h to reduce wear and operating air consumption and better match the desired delivery rate. That also has a reserve capacity of 28 percent available, 6/(4.69 cycles/min) = 1.28, for the usually inevitable future increases needed.

The estimated budget cost will be $4140 plus the 12-in spool cost at 15 percent additional, or $621, = $4761.00. The cost would be $5700 for a 16-in spool with a height of 42 in. Compared to the RAL cost of $2767, the DDV will cost 1.73 times as much. So the evaluation will hinge on wear and maintenance costs and air seal capabilities. For an abrasivelike clinker cement the DDV is probably the best choice.

6.3 Slide Gate Valves

Formerly the most commonly used discharge valve, and in some cases a feeder itself, the slide gate is more often used now for intermittent feed to packaging or for a secondary shutoff above other feeders. When maintenance on the main feeder is necessary, the slide gate provides the shutoff of flow and air leakage.

Where there is not an air seal problem and if the bulk material is free flowing, the slide does act quite well as a feeder. Under a storage hopper, a feed hopper, or a surge hopper feeding a packaging operation, the slide is suitable. Especially when built with adequate slide support surfaces and end seals for free movement and reasonable sealing, it fills the bill at a relatively low cost.

When powered by an air cylinder, a slide gate or valve serves admirably to open and close a feed hopper rapidly, deliver measured batch quantities, and provide dependable service life. Typical costs for a 10-in maintenance slide gate will run $300, and for a process or packaging operation will run up to $800 to $1200. A 12-in maintenance valve can be estimated around $400 and a process slide gate can run up from $900 to $1400 with air cylinder.

6.4 Screw Feeders and Conveyors

Screw feeders or longer conveyors and RALs are probably the two most commonly used discharge devices for dust collectors. Archimedes invented the screw conveyor about 2300 years ago for moving water. Then about 300 years ago the screw conveyor was adapted for moving grain and other bulk materials, and it is still being applied on new materials. Oliver Evans developed a water-wheel-powered flour mill in Wilmington, Delaware, around 300 years ago that used screw and canvas belt conveyors, and bucket elevators to eliminate manual hauling of bags of grain.

Two big advantages of the screw conveyor are the simple operating concept and the relatively low cost. The basic design consists of a spiral screw flight welded on a pipe shaft and supported on bearings in a trough or tube housing. When the shaft is turned, the spiral flights push dust materials along the trough or tube to a discharge opening. As in all volumetric feeders and conveyors, the faster it turns, the more material it delivers.

The capacity of a screw conveyor is determined by its diameter, pitch of the flights, speed of rotation, cross-sectional loading, and the rate at which material is allowed to fall into the screw. Delivery is fairly uniform at higher speeds, but at low speeds the material is more noticeably delivered in pulses as each flight pushes a small pile of material along the trough.

Seal screws

A screw can be used as a pressure seal feeder if it is fully loaded and material is constrained in a tube trough so no leakage of air can occur through the material or around the flights. This application has been called a plug screw or seal screw feeder and has been used with some success. Its volumetric capacity would be calculated on a 100 percent loaded cross section, so it would be roughly 3 times greater than a typical 30 percent loaded screw conveyor.

One critical parameter is the length of the plug section relative to the porosity of the bulk material in the screw so air leakage is controlled; but length is limited by the frictional resistance of the material against the trough and flights. This usually results in a seal or plug section about two to four flights long depending upon the bulk material being conveyed. So the plug screw has a fairly limited area of application and should be thoroughly explored and tested before it is chosen for use.

Screw flight designs

Single-flight standard pitch screws are the most commonly used for cost reasons and have a pitch distance from flight to flight that is equal to the flight diameter. Inclined screws and feeders often use short pitch flights of two-thirds or one-half of the diameter. Vertical screws use a short pitch flight, often with a double flight for an effective pitch of one-fourth of the diameter, at the bottom feed end to start moving material upward. A horizontal feed screw is used to deliver a uniform amount of material into the vertical screw section to provide a full cross-sectional loading. Vertical screws use a round or tubular trough to contain the material in the flights, often tapered at the bottom, to compact the dust and to keep it moving as the flights turn. Multiple-flight screw discharge feeders are often used to ensure positive flow from large hoppers, bins, and silos by pulling material from a larger, or entire, cross section of the hopper or discharge opening.

Twin screw feeders

Twin screw feeders can accomplish the air seal very effectively by using intermeshing screw flights that capture the bulk material in the pockets between the flights. These pockets of material seal against back flow and air leakage, similar to a positive displacement pump. Each screw flight wipes the other flight to keep both clean. An agitator or paddles on a shaft above the twin screws can be driven to break up and help move bulk materials out of the hopper to overcome bridging and rat-holing, and thus help maintain uniform feed rates.

The twin screw feeder can handle many bulk materials that a single-flight screw cannot and does it more consistently than many other feeder types. The pockets of material resist flooding of aerated or fluidized bulk materials and powders and thus control the volumetric rate. Other flight designs can handle sticky or agglomerating materials, such as ribbon flights that are open around the shaft, and help prevent packing in the trough and around the shaft. Twin spiral or ribbon flights that do not intermesh will control the flow of pellets without deforming or degrading them and can also provide some blending or mixing of products in the back flow and tumbling around the ribbon flights in nonintermeshed flights.

Capacity of screw conveyors

Sizing is based on the volume capacity required, similar to RALs. Most horizontal screw conveyors are designed for a 30 percent trough loading, but different material applications may run from 15 to 45 percent cross-sectional loading. The conveyor equipment manufacturers association (CEMA) and most manufacturers of screw conveyors publish an extensive list of materials and their characteristics that can be handled in a screw conveyor which provides the recommended cross-sectional loading and some handling characteristics of each, similar to Tables 6.4 and 6.5. (See also Fig. 6.2.)

Table 6.5 provides the needed *material classification code* data on any bulk material, which can also be used for unlisted but similar materials in determining the volumetric capacity of screw conveyors. Take the first material on the list in Table 6.5, adipic acid, with a density of 45 lb/ft^3 and a material code of 45A100-35. This code gives the average or listed density first, the usual particle size, then the flowability and the abrasiveness number per Table 6.4 definitions, and then any further properties or hazards of the material.

The *component series* provides a guide to selection of conveyor com-

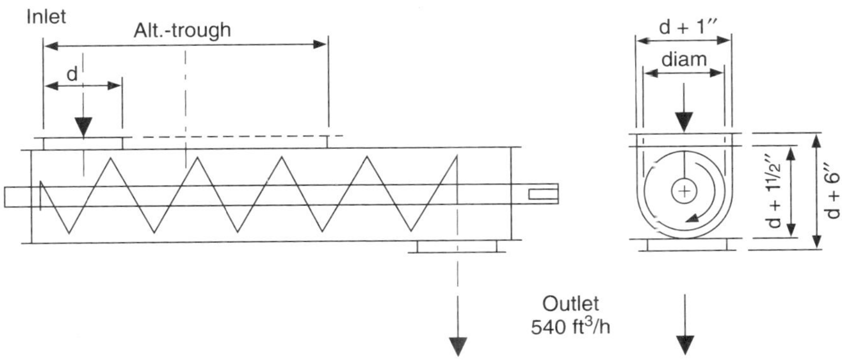

Eq. (6.2): Volume = ft^3/h/rev × rev/min

12″ ϕ: 540 ft^3/h = 12.9 ft^3/h/rev × 41.9 rev/min

14″ ϕ: 540 ft^3/h = 20.8 ft^3/h/rev × 26 rev/min

Figure 6.2 Screw feeders and conveyors.

TABLE 6.4 Material Classification Code Chart

Major class	Material characteristics included			Code designation
Density	Bulk density, loose			Actual pounds per cubic foot
Size		(0.00003937 in)	1 μ	
	Very fine	(0.0003937 in)	10 μ	
	#200 sieve	(0.0029 in)	74 μ	A200
	#100 sieve	(0.0059 in)	159 μ	A100
	#40 sieve	(0.016 in)	3359 μ	A40
	Fine #6 sieve	(0.132 in)		B6
	Granular	½ in and under		C ½
	Granular	3 in and under		D3
	Lumpy	Over 3 in x = actual size		Dx
	Irregular	Stringy, fibrous, cylindrical slabs, etc.		E
Flowability	Very free flow—function over 10			1
	Free flowing—function 4 to 10			2
	Average—function 2 to 4			3
	Sluggish—function less than 2			4
Abrasiveness	Mildly—Index 1–17			5
	Moderately—Index 18–67			6
	Extremely—Index 68–416			7
Miscellaneous properties or hazards	Builds up and hardens			F
	Generates static electricity			G
	Decomposes/deteriorates in storage			H
	Flammability			J
	Plastic or tends to soften			K
	Very dusty			L
	Aerates and becomes fluid			M
	Explosiveness			N
	Sticky, adhesive			O
	Contaminable, affecting use			P
	Degradable, affecting use			Q
	Harmful or toxic fumes or gas			R
	Highly corrosive			S
	Mildly corrosive			T
	Hygroscopic			U
	Interlocks, mats, agglomerates			V
	Oils present			W
	Packs under pressure			X
	Very light and fluffy			Y
	Elevated temperature			Z

TABLE 6.5 Bulk Material Characteristics—Screw Conveyor Application

Material	Density, lb/ft^3	Material code	Component series	Material factor, F_m
Adipic acid	45	45A100-35	2B	0.5
Alfalfa meal	14–22	18B6-45WY	2D	0.6
Alfalfa pellets	41–43	42C1/2-25	2D	0.5
Alfalfa seed	10–15	13B6-15N	1A/B/C	0.4
Almonds, broken	27–30	29C1/2-35Q	2D	0.9
Almonds, whole shelled	28–30	29C1/2-35Q	2D	0.9
Alum, fine	45–50	48B6-35U	1A/B/C	0.6
Alum, lumpy	50–60	55B6-25	2A/B	1.4
Alumina	55–65	58B6-27MY	3D	1.8
Alumina fines	35	35A100-27MY	3D	1.6
Alumina, sized or briquette	65	65D3-37	3D	2.0
Aluminate gel (aluminum hydroxide)	45	45B6-35	2D	1.7
Aluminum chips, dry	7–15	11E-45V	2D	1.2
Aluminum chips	7–15	11E-45V	2D	0.8
Aluminum hydrate	13–20	17C1/2-35	1A/B/C	1.4
Aluminum ore	See Bauxite			
Aluminum oxide	60–120	90A100-17M	3D	1.8
Aluminum silicate	49	49C1/2-35S	3A/B	0.8
Aluminum sulfate	45–58	52C1/2-25	1A/B/C	1.0
Ammonium chloride, crystalline	45–52	49A100-45FRS		
Ammonium nitrate	45–62	54A40-35NTU	3D	1.3
Ammonium sulfate	45–58	52C1/2-35FOTU	1A/B/C	1.0
Antimony powder	—	A100-35	2D	1.6
Apple pomace	15	15C1/2-45Y	2D	1.0
Arsentic oxide (arsenolite)	100–200	110A100-35R		
Arsentic, pulverized	30	30A100-25R	2D	0.8
Asbestos, rock ore	81	81D3-37R	3D	1.2
Asbestos, shredded	20–40	30E-46XY	2D	1.0
Ash, black, ground	105	105B6-35	1A/B/C	2.0
Ashes, coal, dry, ½ in	35–45	40C1/2-46TY	3D	3.0
Ashes, coal, dry, 3 in	35–40	38D3-46T	3D	2.5
Ashes, coal, wet, ½ in	45–50	48C1/2-46T	3D	3.0
Ashes, coal, wet, 3 in	45–50	48D3-46T	3D	4.0
Asphalt, ½ in	45	45C1/2-45	1A/B/C	2.0
Bagasse	7–10	9E-45RVXY	2A/B/C	1.5
Bakelite, fine	30–45	38B6-25	1A/B/C	1.4
Baking powder	40–55	48A100-35	1B	0.6
Baking soda (sodium bicarbonate)	40–55	48A100-25	1B	0.6
Barite (+½–3) (barium sulfate)	120–180	150D3-36	3D	2.6
Barite powder	120–180	150D100-35X	2D	2.0
Barium carbonate	72	72A100-45R	2D	1.6
Bark, wood refuse	10–20	15E-45TVY	3D	2.0
Barley, fine ground	24–36	312B6-35	1A/B/C	0.4

TABLE 6.5 Bulk Material Characteristics—Screw Conveyor Application (*Continued*)

Material	Density, lb/ft^3	Material code	Component series	Material factor, F_m
Barley, malted	31	31C1/2-35	1A/B/C	0.4
Barley, meal	28	28C1/2-35	1A/B/C	0.4
Barley, whole	36–48	42B6-25N	1A/B/C	0.5
Basalt	80–105	93B6-27	3D	1.8
Bauxite, dry, ground	68	68B6-25	2D	1.8
Bauxite, crushed, 3 in	75–85	80D3-36	3D	2.5
Beans, castor, meal	35–40	38B6-35W	1A/B/C	0.8
Beans, whole, shelled	36	36C1/2-15W	1A/B/C	0.5
Beans, navy, dry	48	48C1/2-15	1A/B/C	0.5
Beans, steeped	60	60C1/2-15	1A/B/C	0.8
Bentonite, crude	34–40	37D3-45X	2D	1.2
Bentonite, 100 mesh	50–60	55A100-25MXY	2D	0.7
Benzene, hexachloride	56	56A100-45R	1A/B/C	0.6
Blood, dried	35–45	40D3-45U	2D	2.0
Blood, dried, ground	30	30A100-35U	1A/B	1.0
Bone ash (tri-calcium phosphate)	40–50	45A100-45	1A/B	1.6
Bone black	20–25	23A100-25Y	1A/B	1.5
Bone char	27–40	34B6-35	1A/B	1.6
Bonemeal	50–60	55B6-35	2D	1.7
Bones, ground	50	50B6-35	2D	1.7
Borate of lime	60	60A100-35	1A/B/C	0.6
Borax, fine	45–55	50B6-25T	3D	0.7
Borax, scrng, ½ in	55–60	58C1/2-35	2D	1.5
Boric acid, fine	55	55B6-25T	3D	0.8
Boron	75	75A100-37	2D	1.0
Bran, rice, wheat	16–20	18B6-35NY	1A/B/C	0.5
Braunite (manganese oxide)	120	120A100-36	2D	2.0
Bread crumbs	20–25	23B6-35PQ	1A/B/C	0.6
Brewers grain, dry	14–30	22C1/2-45	1A/B/C	0.5
Brewers grain, wet	55–60	58C1/2-45T	2A/B	0.8
Brick, ground, ⅛	100–120	110B6-37	3D	2.2
Bronze chips	30–50	40B6-45	2D	2.0
Buckwheat	37–42	40B6-25N	1A/B/C	0.4
Calcine flour	75–85	80A100-35	1A/B/C	0.7
Calcium carbide	70–90	80D3-25N	2D	2.0
Calcium carbonate	See limestone			
Calcium fluoride	See fluorspar			
Calcium hydrate and hydroxide	See lime, hydrated			
Calcium lactate	26–29	28D3-45QTR	2A/B	0.6
Calcium oxide	See lime, unslaked			
Calcium phosphate	40–50	45A100-45	1A/B/C	1.6
Calcium sulfate	See gypsum			
Carbon, activated	8–20	14C1/2-16	4B	1.2
Carbon, black, fine	4–6	5A100-35	2A	0.4
Carbon, black, pelleted	20–40	30C1/2	—	1.0
Carborundum	100	100D3-27	3D	3.0
Casein	36	36B6-35	2D	1.6

TABLE 6.5 Bulk Material Characteristics—Screw Conveyor Application *(Continued)*

Material	Density, lb/ft^3	Material code	Component series	Material factor, F_m
Cashew nuts	32–37	35C1/2-45	2D	0.7
Cast-iron chips	130–200	165C1/2-45	2D	4.0
Caustic soda	88	88B6-35RSU	3D	1.8
Caustic soda, flakes	47	47C1/2-45RSUX	3A/B	1.5
Cement, clinker	75–95	85D3-36	3D	1.8
Cement, mortar	133	133B6-35Q	3D	3.0
Cement, Portland	94	94A100-26M	2D	1.4
Cement, Portland, aerated	60–75	68A100-16M	2D	1.4
Chalk, crushed	75–95	85D3-25	2D	1.9
Chalk, pulverized	67–75	71A100-25MXY	2D	1.4
Charcoal, ground	18–28	23A100-45	2D	1.2
Charcoal, lumps	18–28	23D3-45Q	2D	1.4
Chocolate, pressed cake	40–45	43D3-25	2B	1.5
Chrome ore	125–140	133D3-36	3D	2.5
Cinders, blast furnace	57	57D3-36T	3D	1.9
Cinders, coal	40	40D3-36T	3D	1.8
Clay, ceramic, dry, fines	60–80	70A100-35P	1A/B/C	1.5
Clay, calcined	80–100	90B6-36	3D	2.4
Clay, brick, dry, fines	100–120	110C1/2-36	3D	2.0
Clay, dry, lumpy	60–75	68D3-35	2D	1.8
Clover seed	45–48	47B6-25N	1A/B/C	0.4
Coal, anthracite river and culm	55–61	60B6-35TY	2A/B	1.0
Coal, anthracite, sized ½	49–61	55C1/2-25	2A/B	1.0
Coal, bituminous, mined	40–60	50D3-35LNXY	1A/B	0.9
Coal, bituminous, sized	45–50	48D3-35QV	1A/B	1.0
Coal, bituminous, slack	43–50	47C1/2-45T	2A/B	0.9
Coal, lignite	37–45	41D3-35T	2D	1.0
Cocoa beans	30–45	38C1/2-25Q	1A/B	0.5
Cocoa, powdered	30–35	33A100-45XY	1B	0.9
Coconut, shredded	20–22	21E-45	2B	1.5
Coffee, chaff	20	20B6-25MY	1A/B	1.0
Coffee, green beans	25–32	29C1/2-25PQ	1A/B	0.5
Coffee, ground, dry	25	25A40-35P	1A/B	0.6
Coffee, ground, wet	35–45	40A40-45X	1A/B	0.6
Coffee, roast beans	20–30	25C1/2-25PQ	1B	0.4
Coffee, soluble	19	19A40-35PUY	1B	0.4
Coke, breeze	25–35	30C1/2-37	3D	1.2
Coke, loose	23–35	30D7-37	3D	1.2
Coke, petrol	35–45	40D7-37	3D	1.3
Compost	30–50	40D7-45TV	3A/B	1.0
Concrete premix	85–120	103C1/2-36U	3D	3.0
Copper ore	120–150	135Dx-36	3D	4.0
Copper ore, crushed	100–150	125D3-36	3D	4.0
Copper sulfate (bluestone)	75–95	85C1/2-35S	2A/B/C	1.0
Copra, ground cake	40–45	43B6-445HW	1A/B/C	0.7
Copra, lumpy cake	25–30	28D3-35HW	2A/B/C	0.8
Copra, lumpy	22	22E-35HW	2A/B/C	1.0
Copra, meal	40–45	42B6-35HW	2D	0.7

TABLE 6.5 Bulk Material Characteristics—Screw Conveyor Application (*Continued*)

Material	Density, lb/ft^3	Material code	Component series	Material factor, F_m
Cork, fine, ground	5–15	10B6-35JNY	1A/B/C	0.5
Cork, granulated	12–15	14C1/2-35JY	1A/B/C	0.5
Corn, cracked	40–50	45B6-25P	1A/B/C	0.7
Corncobs, ground	17	17C1/2-25Y	1A/B/C	0.6
Corn ear	56	56E-35	2A/B	
Corn germ	21	21B6-35PY	1A/B/C	0.4
Corn grits	40–45	43B6-35P	1A/B/C	0.5
Cornmeal	32–40	36B6-35P	1A/B	0.5
Corn oil cake	25	25D7-45HW	1A/B	0.6
Corn seed	45	445C1/2-25PQ	1A/B/C	0.4
Corn sugar	30–35	33B6-35PU	1B	1.0
Cottonseed cake, crushed	40–45	43C1/2-45HW	1A/B	1.0
Cottonseed, dry, delinted	22–40	31C1/2-25X	1A/B	0.6
Cryolite dust	75–90	83A100-36L	2D	2.0
Cullet, fine	80–120	100C1/2-37	3D	2.0
Diatomaceous earth	11–17	14A40-36Y	3D	1.6
Dicalcium phosphate	40–50	45A40-35	1A/B/C	1.6
Disodium phosphate	25–31	28A40-35	3D	0.5
Distillers grain, spent dry	30	30B6-35	2D	0.5
Dolomite, crushed	80–100	90C1/2-36	2D	2.0
Earth loam, dry, loose	76	76C1/2	2D	1.2
Ebonite, crushed	63–70	67C1/2-35	1A/B/C	0.8
Egg powder	16	16A40-35MPY	1B	1.0
Epsom salts (magnesium sulfate)	40–50	45A40-35U	1A/B/C	0.8
Feldspar, ground	65–80	73A100-37	2D	2.0
Feldspar, powder	100	100A200-36	2D	2.0
Ferrous sulfide, 100M powder	105–120	113A100-36	1A/B/C	2.0
Ferrous sulphate	50–75	63C1/2-35U	2D	1.0
Fish meal	335–340	38C1/2-45HP	1A/B/C	1.0
Flaxseed	43–45	44B6-35X	1A/B/C	0.4
Flour wheat	33–40	37A40-45LP	1B	0.6
Flue dust from basic oxygen furnace	45–60	53A40-36LM	3D	3.5
Fluorspar, fine (calcium fluoride)	80–100	90B6-36	2D	2.0
Fly ash	30–45	38A40-36M	3D	2.0
Fullers earth, dry, raw	30–40	35A40-25	2D	2.0
Fullers earth, oily, spent	60–65	63C1/2-45OW	3D	2.0
Gilsonite	37	37C1/2-35	3D	1.5
Glass batch	80–100	90C1/2-37	3D	2.5
Glue, ground	40	40B6-45U	2D	1.7
Granite, fine	80–90	85C1/2-27	3D	2.5
Graphite flake	40	40B6-25LP	1A/B/C	0.5
Graphite flour	28	28A100-35LMP	1A/B/C	0.5
Guano, dry	70	70C1/2-35	3A/B	2.0
Gypsum, calcined	55–60	58B6-35U	2D	1.6
Gypsum, calcined, powder	60–80	70A100-35U	2D	2.0

TABLE 6.5 Bulk Material Characteristics—Screw Conveyor Application *(Continued)*

Material	Density, lb/ft^3	Material code	Component series	Material factor, F_m
Hops, spent, dry	35	35D3-35	2A/B/C	1.0
Ice, crushed	35–45	40D3-35O	2A/B	0.4
Ilmenite ore	140–160	150D3-37	3D	2.0
Iron ore, concentrated	120–180	150A40-37	3D	2.2
Iron oxide pigment	25	25A100-36LMP	1A/B/C	1.0
Kaolin clay, talc	42–56	49A40-35LMP	2D	2.0
Lactose	32	32A40-35PU	1B	0.6
Lead arsenate	72	72A40-35R	1A/B/C	1.4
Lead carbonate	240–260	250A40-35R	2D	1.0
Lead oxide, red	30–180	105A200-35LP	2D	1.2
Lime, ground, unslaked	60–65	63B6-35U	1A/B/C	0.6
Lime, hydrated	40	40B6-35LM	2D	0.8
Limestone dust	55–95	75A40-46MY	2D	1.6–2.0
Lithopne	45–50	48A325-35MR	1A/B	1.0
Malt, dry, ground	20–30	25B6-35NP	1A/B/C	0.5
Magnesium chloride	33	33C1/2-45	1A/B	1.0
Manganese dioxide	70–85	78A100-35NRT	2A/B	1.5
Manganese oxide	120	120A100-36	2D	2.0
Manganese sulfate	70	70C1/2-37	3D	2.4
Marble, crushed	80–95	88B6-37	3D	2.0
Marl clay	80	80Dx-36	2D	1.6
Mica, flakes	17–22	20B6-16MY	2D	1.0
Mica, pulverized	13–15	14A100-36M	2D	1.0
Milk powder	20–45	33B6-25PM	1B	0.5
Milo, ground	32–36	34B6-25	1A/B/C	0.5
Molybdenite powder	107	107B6-26	2D	1.5
Monosodium phosphate	50	50B6-36	1D	0.6
Naphthalene flakes	45	445B6-35	1A/B/C	0.7
Niacin	35	35A40-35P	2D	0.8
Oats flour	35	35A100-35	1A/B/C	0.5
Oxalic acid, crystals	60	60B6-35QS	1A/B	1.0
Oyster shell, grounds	50–60	55C1/2-36T	3D	1.6–2.0
Paper pulp, 4%	62	62E-45	2A/B	1.5
Peanut meal	30	30B6-35P	1B	0.6
Perlite, expanded	8–12	10C1/2-36	2D	0.6
Phosphate acid fertilizer	60	60B6-25T	2A/B	1.4
Phosphate rock, pulverized	60	60B6-36	2D	1.7
Polystyrene beads	40	40B6-35PQ	1B	0.4
Polyvinyl chloride, powder	20–30	25A100-45KT	2B	1.0
Polyethylene, resin pellets	30–35	33C1/2-45Q	1A/B	0.4
Potash, dry	70	70B6-37	3D	2.0
Potassium carbonate	51	51B6-36	2D	1.0
Potassium nitrate	80	80B6-26NT	2D	1.2
Potassium sulfate	42–48	45B6-46X	2D	1.0
Potato flour	48	48A2300-35MNP	1A/B	0.5
Pyrite, pellets	120–130	125C1/2-26	3D	2.0
Quartz	70–80	75A100-27	3D	1.7
Rice bran	20	20B6-35NY	1A/B/C	0.4
Rice, hulled	45–49	47C1/2-25P	1A/B/C	0.4

TABLE 6.5 Bulk Material Characteristics—Screw Conveyor Application (*Continued*)

Material	Density, lb/ft^3	Material code	Component series	Material factor, F_m
Rosin	65–68	67C1/2-45Q	1A/B/C	1.5
Rubber recl., ground	23–50	37C1/2-45	1A/B/C	0.8
Rubber, pellet	50–55	53D3-45	2A/B/C	1.5
Rye	42–48	46B6-15N	1A/B/C	0.4
Safflower	50	50D3-26	2D	0.6
Salt cake, dry (sodium sulfate; pulverized)	65–85	75B6-36TU	3D	1.7
Salicylic acid	29	29B6-37U	3D	0.8
Salt, dry, fine	70–80	75B6-36TU	3D	1.7
Sand, dry (silica)	90–100	95B6-27	3D	2.0
Sand, foundry	90–100	95D3-37Z	3D	2.6
Sand, zircon, resin-coated	115	115A100-27	3D	2.3
Sawdust, dry	10–13	12B6-45UX	1A/B/C	0.7
Sea coal	65	65B6-36	2D	1.0
Sesame seed	27–41	34B6-26	2D	0.6
Shale, crushed	85–90	88C1/2-362D	2D	2.0
Shellac, powder	31	31B6-35P	1B	0.6
Silica flour	80	80A40-46	2D	1.5
Slag furnace	60–65	63C1/2-37	3D	2.2
Slate, ground	82–85	84B6-367	2D	1.6
Sludge sewage, dry	40–50	45E-47TW	3D	0.8
Sludge sewage, ground	45–55	50B6-46S	2D	0.8
Soap, beads/granules	15–35	25B6-35Q	1A/B/C	0.6
Soap detergent	15–50	33B6-35FQ	1A/B/C	0.8
Soapstone, talc	40–50	45A200-45XY	1A/B/C	2.0
Soda ash, heavy	55–65	60B6-36	2D	1.0
Soda ash, light	20–35	28A40-36Y	2D	0.8
Sodium aluminate, ground	72	72B6-36	2D	1.0
Sodium aluminum sulfate	75	75A100-36	2D	1.0
Sodium nitrate	70–80	75D3-25NS	2A/B	1.2
Sodium phosphate	50–60	55A-35	1A/B	0.9
Sodium sulfite	96	96B6-46X	2D	1.5
Soybean cake	40–43	42D3-35W	1A/B/C	1.0
Soybean flour	27–30	29A40-35MN	1A/B/C	0.8
Starch	25–50	38A40-15M	1A/B/C	1.0
Steel turnings	100–150	125D3-46WV	3D	3.0
Sugar beet, pulp, dry	12–15	14C1/2-26	2D	0.9
Sugar beet, pulp, wet	25–45	35C1/2-35X	1A/B/C	1.2
Sugar, refined, granulated, dry	50–55	53B6-35PU	1B	1.0–1.2
Sugar, R.G., wet	55–65	60C1/2-35X	1B	1.4–2.0
Sugar, powdered	50–60	55A100-35PX	1B	0.8
Sugar, raw	55–65	60B6-35PX	1B	1.5
Sulfur, crushed	50–60	55C1/2-35N	1A/B	0.8
Sulfur, lumpy	80–85	83D3-35N	2A/B	0.8
Sulfur, powdered	50–60	55A40-35MN	1A/B	0.6
Sunflower seed	19–38	29C1/2-15	1A/B/C	0.5
Talcum, ½ in	80–90	85C1/2-36	2D	0.9

TABLE 6.5 Bulk Material Characteristics—Screw Conveyor Application (*Continued*)

Material	Density, lb/ft^3	Material code	Component series	Material factor, F_m
Talcum powder	50–60	55A200-36M	2D	0.8
Tanbark, ground	55	55B6-45	1A/B/C	0.7
Titanium dioxide	140	140A-27	4C	2.0
Tobacco scraps	15–25	20D3-45Y	2A/B	0.8
Tobacco snuff	30	30B6-45MQ	1A/B/C	0.9
Tricalcium phosphate	40–50	45A40-45	1A/B	1.6
Triple sugar phosphate	50–55	53B6-36RS	3D	2.0
Trisodium phosphate	60	60C1/2-36	2D	1.7
Trisodium phosphate, granular	60	60B6-36	2D	1.7
Trisodium phosphate, pulverized	50	50A40-36	2D	1.6
Tung nut meat	28	28D3-25W	2A/B	0.8
Tung nuts	25–30	28D3-15	2A/B	0.7
Urea polls, coated	43–46	45B6-25	1A/B/C	1.2
Vermiculite, exposed	16	16C1/2-35Y	1A/B	0.5
Vermiculite ore	80	80D3-36	2D	1.0
Vetch	46	46B6-16N	1A/B/C	0.4
Walnut shells, crushed	35–45	40B6-36	2D	1.0
Wheat	45–48	47C1/2-25N	1A/B/C	0.4
Wheat, cracked	40–45	43B6-25N	1A/B/C	0.4
Wheat germ	18–28	23B6-25	1A/B/C	0.4
White lead, dry	75–100	88A40-36MR	2D	1.0
Wood chips, Scrnd	10–30	20D3-45VY	2A/B	0.6
Wood flour	16–36	26B6-35N	1A/B	0.4
Wood shavings	8–16	12E-45VY	2A/B	1.5
Zinc concentrate, residue	75–80	78B6-37	3D	1.0
Zinc oxide, heavy	30–35	33A100-45X	1A/B	1.0
Zinc oxide, light	10–15	13A100-45XY	1A/B	1.0

ponents, such as bearings from manufacturer's catalogs and proposals. The *material factor* is used in calculating the drive horsepower required as discussed in manufacturer's catalogs also.

A typical material would be the first in Table 6.5, adipic acid, with a listed density of 45 lb/ft^3, material code of 45A100-35, component series 2B, and material factor F_m = 0.5. This material code gives the listed, or an average density, which for adipic acid is a density of 45 lb/ft^3, a typical particle size A100 of very fine at 100 sieve or 0.0059 in (159 μ), and the 35 indicates an average flowability (3) and mild abrasiveness (5). The letter designation of size, A through E, and the numbers for flowability and abrasiveness, 1 through 7, determine the degree of trough loading, in this case 30%-A cross-sectional loading by reference to Table 6.6 under material code. So for a 12-in-diameter screw, the maximum recommended speed is 90 rpm, and the capacity at maximum revolutions per minute is 1160 ft^3/h, or the capacity at 1 rpm is 12.9 ft^3/h.

Referring to Table 6.6 we find the A-35 code number in the 30%-A

TABLE 6.6 Horizontal Screw Conveyor Capacity

Material code (from Table 6.4)	Trough cross-sectional load	Screw diameter, in	Maximum recommended speed, rpm	Capacity, ft^3/h	
				at max. rpm	at 1 rpm
A-15	45%	6	165	368	2.23
A-25	45%	9	155	1,270	8.2
B-15	45%	10	150	1,710	11.4
B-25	45%	12	145	2,820	19.4
C-15	45%	14	140	4,370	31.2
C-25	45%	16	130	6,060	46.7
	45%	18	120	8,120	67.6
	45%	20	110	10,300	93.7
	45%	24	100	16,400	164.0
A-35 E-15	30%-A	6	120	180	1.49
A-45 E-25	30%-A	9	100	545	5.45
B-35 E-35	30%-A	10	95	720	7.57
B-45 E-45	30%-A	12	90	1,160	12.9
C-35	30%-A	14	85	1,770	20.8
C-45	30%-A	16	80	2,500	31.2
D-15	30%-A	18	75	3,380	45.0
D-25	30%-A	20	70	4,370	62.5
D-45	30%-A	24	65	7,100	109.0
A-16 D-16	30%-B	6	60	90	1.49
A-26 D-26	30%-B	9	55	300	5.45
A-36 D-36	30%-B	10	55	420	7.6
A-46 D-46	30%-B	12	50	645	12.9
B-16 E-16	30%-B	14	50	1,040	20.8
B-26 E-26	30%-B	16	45	1,400	31.2
B-36 E-36	30%-B	18	45	2,025	45.0
B-46 E-46	30%-B	20	40	2,500	62.5
C-16/C-46	30%-B	24	40	4,360	109.0
A-17 D-17	15%	6	60	45	0.75
A-27 D-27	15%	9	55	150	2.72
A-37 D-37	15%	10	55	210	3.8
A-47 D-47	15%	12	50	325	6.46
B-17 E-17	15%	14	50	520	10.4
B-27 E-27	15%	16	45	700	15.6
B-37 D-37	15%	18	45	1,010	22.5
B-47 D-47	15%	20	40	1,250	31.2
C-17/C-47	15%	24	40	2,180	54.6

section for the degree of cross-sectional loading. This tells us we could use a 6-in screw to deliver a volume at 1 rpm of 1.49 ft^3/h, a 9-in at 5.45, a 10-in at 7.57, a 12-in at 12.9 $ft^3/h \cdot rev$, and the larger sizes as listed. A screw conveyor to receive the required volume of 7.82 ft^3/min from the RAL and DDV delivery rate discussed above will need the following volume.

Screw conveyor volumetric capacity. Referring to Table 6.6 select the diameter or size, or try several sizes, of screw conveyor that will deliver the necessary volume at around 40 rpm. This should start with the capacity delivered per revolution per minute. Then multiply that diameter's capacity at 1 rpm by 40 rpm which is the recommended speed for wear and for economical drive costs. That basic capacity in ft^3/h is for a horizontal conveyor or feeder, with no supply or delivery flow problems before or after the screw that may cause a bottleneck to flow.

For adequate capacity to handle an upstream feeder like a RAL or a DDV, the downstream capacity should be increased by about 15 percent to prevent bottlenecks to the upstream flow, especially from surges. Experience will guide the actual increase needed, but around 15 percent is a generally safe allowance.

From the cement RAL example above the required capacity was 7.82 ft^3/min, so an equivalent capacity screw conveyor would be:

$$\text{Screw conveyor capacity} = (\text{ft}^3/\text{h} \cdot \text{rev}) \times (\text{rpm}) \qquad (6.3)$$

$$469 \text{ ft}^3/\text{h} = 60 \text{ min/h} \times 7.82 \text{ ft}^3/\text{min} = (\text{ft}^3/\text{h} \cdot \text{rev}) \times (\text{rpm})$$

$$= 12.9 \text{ ft}^3/\text{h} \cdot \text{rev} \times 36.4 \text{ rpm} = 469.56 \text{ ft}^3/\text{h}$$

A 12-in-diameter screw for A-16 or A-26 aerated portland cement at 80 lb/ft^3 could supply the capacity at 36.4 rpm. For a 15 percent additional capacity allowance, the speed needed would be 469 $ft^3/h \times 1.15/12.9$ $ft^3/h \cdot rev = 41.8$ rpm. This is a reasonable and economical drive speed to deliver 539 ft^3/h of 80 lb/ft^3 portland cement.

The first adequate size at the Table 6.6 30%-B cross-sectional loading for A-16, A-26, A-36 cement is the 12-in diameter at 12.9 $ft^3/h/rev$, and next is the 14-in diameter at 20.8 $ft^3/h \cdot rev$.

$$\text{Speed for 14-in-diameter screw} = 20.8 \text{ ft}^3/\text{h} \cdot \text{rev} \times \text{rpm} = 539 \text{ ft}^3/\text{h}.$$

Speed for 539 $ft^3/h = 20.8 \times 26$ rpm, which is certainly conservative, and good for lower wear, but probably at higher cost for the speed reduction and the 14-in conveyor size.

Recommended 40 rpm operating speed

For long service and low wear, plus economy with standard drive components, a recommended operating speed on horizontal screw conveyors is around 40 rpm. Thus solving Eq. (6.3) for capacity per hour:

$$\text{Actual sizing capacity} = 40 \text{ rpm} \times 13.5 \text{ ft}^3/\text{h} \cdot \text{rev}$$

$$= 540 \text{ ft}^3/\text{h}$$

A speed close to the recommended nominal 40 rpm can easily be achieved by varying the V-belt drive speed reduction. In this case there are two good choices, depending on project considerations. For lowest cost and minimum height, the 12-in diameter is an excellent choice with reserve capacity for the future still available by speeding up the V-belt drive. The 14-in diameter is also excellent for reduced wear at the lower speed with more than adequate reserve for the future.

Drive speed calculations. Using the most economical drive motors at 1750 rpm and gear reducers of 15:1, 20:1, or 25:1 reduction ratio as most commonly used on screw conveyor drives,

$$\text{Drive output speed} = \frac{\text{motor rpm}}{\text{reducer ratio} \times \text{V-belt ratio}} \tag{6.4}$$

Thus, a typical speed range for a 20:1 reducer and reasonable V-belt ratios is:

$$1750 \text{ rpm} \times 1/20 \times 1/2.188 = 40 \text{ rpm}$$

$$1750 \text{ rpm} \times 1/20 \times 1/1.5 = 58.3 \text{ rpm}$$

$$1750 \text{ rpm} \times 1/20 \times 1/3.364 = 26 \text{ rpm}$$

Normal V-belt pulley or sheave ratios can range from 1.0 to 3.5 so these are acceptable ratios. The calculated operating speed of 41.8 rpm is acceptable for normal wear and capacity in the cement case above.

For the low wear on the screw and trough, and if there is no space problem, the 14-in diameter would probably be the best choice operating at 26 rpm with a 25:1 reducer ratio. Economics always enter the picture and would dictate use of the 12-in diameter screw operating at 41.8 rpm if either space or cost is critical.

When bidding a competitive job, the base bid should be on the 12-in diameter with an option offered on the 14-in diameter for the reasons mentioned.

For clinker cement the material code is 85D3-36 in Table 6.5, indicating an average density of 85 lb/ft^3, a code designation of D3 indicating a granular particle size of 3 in and under, and 36 indicating an average flowability (3) and a moderately abrasive (6) product. As stated earlier the customer has specified a density of 80 lb/ft^3 which we used in the volume capacity calculation.

Note that actual product specifications change and there are always new products to be analyzed; thus Table 6.5 is a guide only. Test new products or critical cases where the specifications and characteristics are at all in doubt. But these values can be used as a guide and reference for similar products in preliminary planning. Quotations should be obtained on each screw feeder and conveyor project and the performance firmly guaranteed.

Mass flow conveyors

The application of mass flow conveyors is similar to screw conveyors, since most use a screw conveyor trough of U-shape or rectangular cross section. A chain pulls flights through the trough to push bulk material along the trough. Trough covers, inlets, and outlets are similar to standard screw designs, and this conveyor can be made dust-tight like screw conveyors. (See Fig. 6.3.)

U-trough mass flow conveyor capacity is around 2 to 6 times greater than a 30 percent loaded screw for the same trough size, and twice that of a 100 percent loaded screw feeder. The cross-sectional

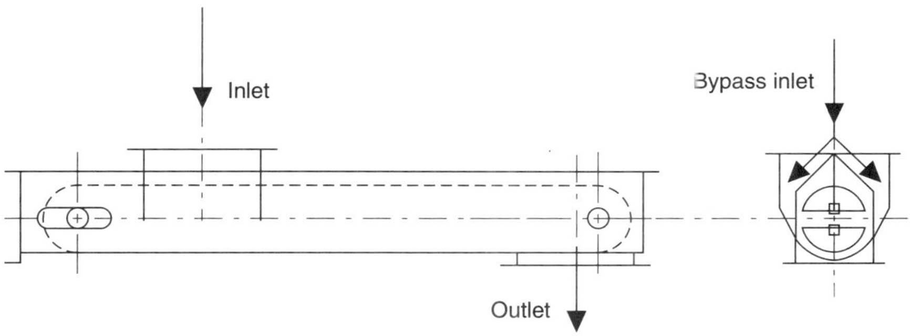

Figure 6.3 Mass flow feeders and conveyors.

TABLE 6.7 Mass Flow Conveyor Capacities—Screw Troughs

Trough diameter, in	100-ft/min speed			150 ft/min speed		
	Capacity, ft^3/h	Drive, rpm	Power, hp/ft length	Capacity, ft^3/h	Drive, rpm	Power, hp/ft length
9	2,100	61	0.046	3,100	92	0.060
12	3,400	45	0.060	5,100	68	0.080
14	4,800	38	0.075	7,200	57	0.103
16	6,400	35	0.089	9,700	53	0.123
18	8,500	30	0.106	12,800	46	0.148
20	10,700	27	0.132	16,000	41	0.189
24	15,400	23	0.168	23,000	35	0.242

Note: Capacity and power are based on free-flowing small grains weighing 48 lb/ft^3.

load is greater and the operating speed can be 1.5 to 3 times faster than a screw as measured in feet per minute of bulk material travel. Horsepower is about half that of a screw of the same length and material capacity since material friction on the screw flight is eliminated. Lengths up to 120 ft are typically used. Sizes, capacities, and horsepower of mass flow conveyors in trough diameter sizes from 9 to 24 in and for operating speeds from 100 to 150 ft/min are listed in Table 6.7.

Multiple inlets and extended hopper fed inlets can be handled with a bypass shroud at the inlet. This shroud has a V-shaped cover over the center of the chain and flights to bypass the incoming dust around the moving parts, as well as to allow controlling the volumetric or cross-sectional loading by a predetermined or adjustably set inlet flow passage size or slide valve.

Corrosive, abrasive, and higher-temperature materials should be handled as much as possible with the trough loading low enough to pass under the chain and intermediate bearings. Idler supports are used on long conveyors to hold the return chain above the lower chain and flights with the intermediate bearings mounted either inside or outside the trough depending upon the nature of the materials.

Ultra-high molecular weight plastic (UHMW) flights and trough liners can be used to reduce friction, wear, and corrosion with many bulk solid materials. Stainless steel, abrasion-resistant steel, other metals, and Teflon can be used also. Cost is about 25 percent higher than a screw conveyor of the same trough size. But since the drag or mass flow conveyor provides at least twice the capacity for the same size, the cost per cubic foot per minute is about two-thirds that of a screw conveyor.

TABLE 6.8 Friable Material Handled by Aero Conveyor

Material	Bulk density, lb/ft^3	Conveying rate, lb/h	Distance, ft	Power, hp	Degradation percent of fines increase
Cut tea	20	12,000	60	3	Below 1%
Coffee bean	38	40,000	100	5	Below 1%
Silica gel	43	9,000	58	3	Below 1%
Cereal flakes	10	6,000	67	3	2.7%
Dried onion flakes	19	20,000	70	3	Negligible
Garlic powder	40	44,000	30	3	Negligible
Silver nitrate	160	2,000	15	3	Below 1%

Aero conveyor—drag and fluidized

This disk and cable conveyor runs in 3- or 4-in-diameter tubing. The cable is pulled by a mechanical drive and no separate air mover is required. It can typically convey up to 120-ft lengths and run at 200 to 500 ft/min. The high travel speed causes aeration of the dust, and the round flights pull the dust mass along the tubular trough. It will handle 30 to 60 percent of the cross-sectional volume of the tube or pipe versus about 20 to 30 percent for a dense phase pneumatic conveyor. Horsepower is about one-half that of a pneumatic conveyor, or 3 hp versus up to 7.5 hp for a pneumatic (see Table 6.8). The cost should be about the same as the mass flow conveyor.

6.5 Belt Conveyors

Belt conveyors and belt feeders have been in use for several hundred years, since the early days of cotton and leather belts. These were used in grain mills, cotton gins, and of course the early line shaft drives for power transmission.

Flat or troughed belt dischargers are used under silos and bins, generally with a wide and long opening up to 4 or 5 ft wide and 10 to 20 ft long. However, they do not normally provide a good air seal, so they are not often used under dust collectors by themselves. (See Fig. 6.4.)

Where there is a bin activator or a valve to provide an air seal, to induce flow out of a hopper with some uniformity, and generally to provide some initial feed rate control, belt conveyors can be very suit-

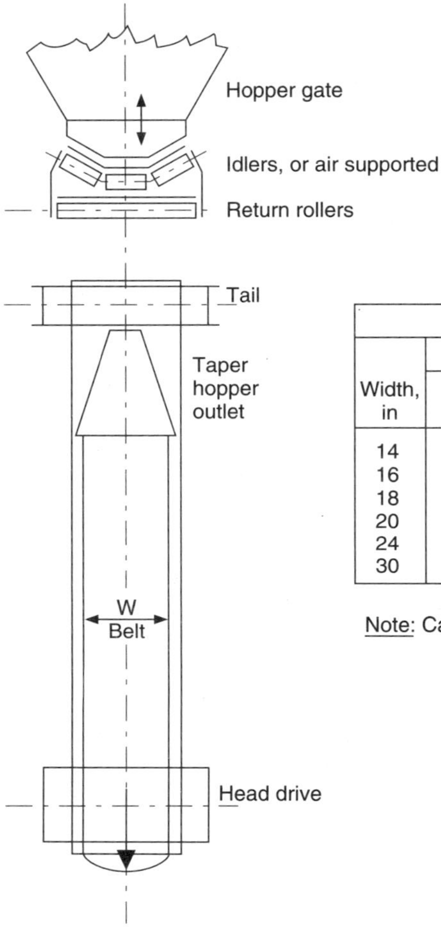

Capacity TPH—100 ft/min						
	Bulk density, lb/ft^3				Max. ft/min	
Width, in	30	50	100	150	Light free	Med. large
14	10	17	34	51	400	250
16	13	22	44	66	500	300
18	17	28	56	84	500	300
20	20	34	68	102	600	400
24	30	50	100	150	600	400
30	47	79	158	237	750	500

Note: Capacity is directly proportional to belt speed.

Figure 6.4 Belt feeders, conveyors, and capacities.

able. They have low-wear, low-maintenance, low-power requirements, and low cost for transportation over a distance of more than 5 or 10 ft. Some careful guidelines for application are

- A tapered discharge slot hopper opening in the plan view, narrow at the beginning or tail pulley end, and wider at the delivery or drive end. This provides for hopper discharge along the entire length and generally helps bulk solids flow out of the bin or hopper.
- A discharge end vertical gate to set the volumetric flow opening. This should not be used for shutoff of flow since it will jam the bulk material behind the gate and can cause compaction.

- Belt speed should be designed to provide the basic discharge flow rate, and the end gate used only for trimming the volume rate. Shutoff is normally accomplished by turning the drive motor off.
- Adequate head pulley diameter and traction will be needed to pull the fully loaded belt. Adequate bed support for the belt under a full head of material will be needed.
- Slider bed design is usually best for a short belt discharge up to 20 ft long. A flat or troughed slider bed prevents the spillage of material that occurs over a roller bed if the belt sags too much between idler rolls.
- Side skirts with flexible seal and guide strips are recommended on both sides of the belt to contain the material, prevent spillage, and help seal the discharge. (See Fig. 6.4.)

6.6 Vibrating Bin Dischargers

This is a flow-aiding device and volumetric feeder combined. The design is based on a truncated hopper or bin bottom discharge at half the full bin diameter or less, fitted with a vibrating shallow cone that is gyrated horizontally in a circular pattern. The discharge cone and outer fixed housing are connected by a flexible sleeve to the main bin to seal the flow path and absorb the gyratory motion. The discharger is supported on short movable arms from the main bin or hopper bottom.

Flow is induced by starting the gyratory motion of the inner cone which pushes and then releases the stored bulk solid dust. This tends to break up any bridging or rat-holing and lets the loosened dust fall through the circular gap between the cone and the main bin walls. The gyratory housing has a discharge opening sized for the required flow rate, typically 6- to 14-in diameter. The inner cone itself has been found to enhance bulk material flow by relieving the pressure of unrestricted center flow.

The amount of gyratory motion can be designed, and adjusted, for the range of flow desired. Obviously an on-off control of the eccentric drive can start and stop the material flow. One design has a secondary cone mounted below the flow-aiding primary cone, with the intent of further aiding the flow by breaking up any compaction in the lower chamber. Another design has a ladder grate below the primary cone for the same purpose. Connection to a downstream conveyor would be made with a flexible sleeve or skirts to keep from transmitting the gyratory or vibrating motion.

The inherent advantage of a vibrating bin activator or discharger is overcoming or aiding the flow of difficult bulk solids. Its cost is around 50 percent higher than a double dump valve, so it is 3 to 4 times that of a RAL. But it has these advantages:

- Overcomes many difficult flow problems
- Needs very low maintenance and reduces abrasive wear
- Can reduce overall hopper height, and/or increase storage volume in the same plant height available
- Remixes dust particle sizes that may have been segregated during filling and storage
- Handles a large head of material in the bin above, such as encountered in silos
- Can be readily used in a loss-in-weight system where the bin, hopper, or silo is weighed continuously by load cells or strain gauges

Readily available sizes are from 2 to 18 ft in diameter. Recently some square units, and an optional lifting and lowering actuator for the gyratory cone to provide shutoff flow control, have been developed. The cone lift may also be set at a selected height in its vertical travel to control the rate of flow. (See Fig. 6.5.)

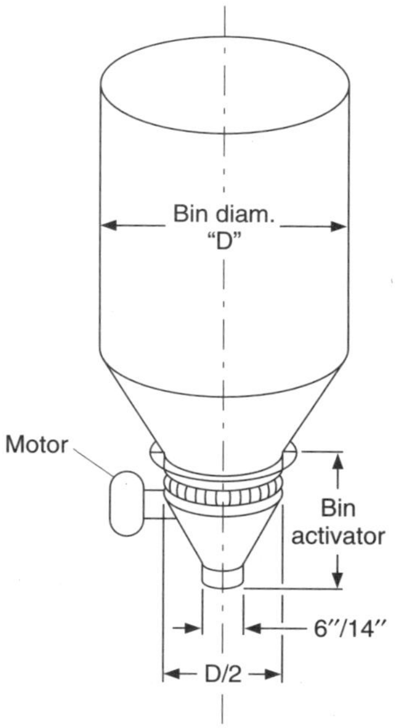

Figure 6.5 Vibrating bin-hopper-silo discharger. Typically the bin activator is sized at one-half the bin diameter. The activator is driven in an oscillating and vibrating motion, and can feed conveyors, feeders, weigh hoppers, or process vessels. The activator can be from 2 to 18 ft in diameter, and the discharge outlet can be from 6 to 14 in or larger. Flow rates can be as high as 5000 ft^3/h, and feed a 48-in-wide belt conveyor. Feed rate accuracy can be held to (±) 1 to 2 percent. They can also solve the problem of inducing balky bulk solids to flow out of a bin or hopper.

6.7 Other Discharge Devices

Many other discharge feeders and devices are used and may be the right one for a particular product. Some industries have found by experience that certain types are most suitable. But keep on the lookout for new developments and applications! Some of these are important and widely used discharge devices or are used in conjunction with other devices:

- Slide gates are very important flow controls especially for manual, timed, or sequenced control of bulk flow. Typical uses are for maintenance shutoff before or after another type flow valve, and measured discharge on each cycle by time or degree of opening for dust or powder feed on each press stroke, bag filling, or discharge of a dust collector.
- Flexible tube discharge gate that opens when a sufficient head of material has built up above the flexible tube.
- For reduced wear the flexible iris type valve, dome or disk valves, the flexible tube above, and others are available. A recent development is the inflatable seal on butterfly, slide, and other valves to provide an airtight seal when inflated, but an easy opening and low-wear seat when deflated for opening.

6.8 Fluidized Feeders, Conveyors, and Capacities

Fluidization has long provided a useful conveyor and feeder method with very low abrasive wear and maintenance. The only equipment that usually requires maintenance is the blower or compressed air supply for fluidization. The porous fluidizing medium may need replacing every 5 to 10 years of service. The fluidized conveyor has been widely used for finely powdered and granular bulk materials such as portland cement, flour, pulverized coal, bentonite and other clays, silica flour, fluorspar, phosphate, soda ash, vacuum dried salt, sodium sulfate, ground bauxite, alumina, magnesite, and uranium dioxide. Sizes of fluidized conveyors and feeders from 4- to 16-in width can handle capacities from 300 to 8000 ft^3/h (see Fig. 6.6).

The only power required is the centrifugal or positive displacement blower, or a pressure-regulated compressed air supply. The quantity of free air required depends upon the characteristics of the bulk material or powder, the downward slope of the conveyor (usually within a range of 2° to 10°), depth of the powder layer, width and length and thus area of the fluidization surface, permeability of the medium, and

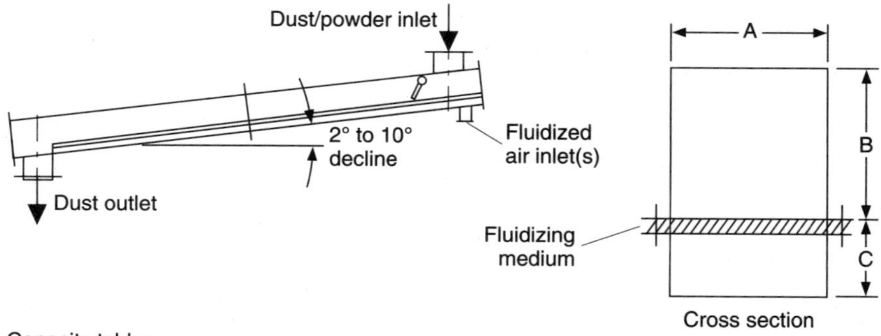

Capacity table:

Width	4″	6″	8″	10″	12″	14″	16″
A	4 1/8	6 1/8	8 1/8	10 1/8	12 1/8	14 1/8	16 1/8
B	4	4	6	6	8	10	10
C	2	2	3	3	3	3	3
CFH	300	600	1000	2000	3000	4000	8000
65 lb/ft^3 TPH	9.75	19.5	32.5	65.0	97.5	130.0	260.0
20 lb/ft^3 TPH	3	6	10	20	30	40	80
Wt/ft	7 lb/3.2 kg	9/4.1	11/5	14/6.4	16/7.3	18/8.2	21/9.6

Figure 6.6 Fluidized feeders, conveyors, and capacities.

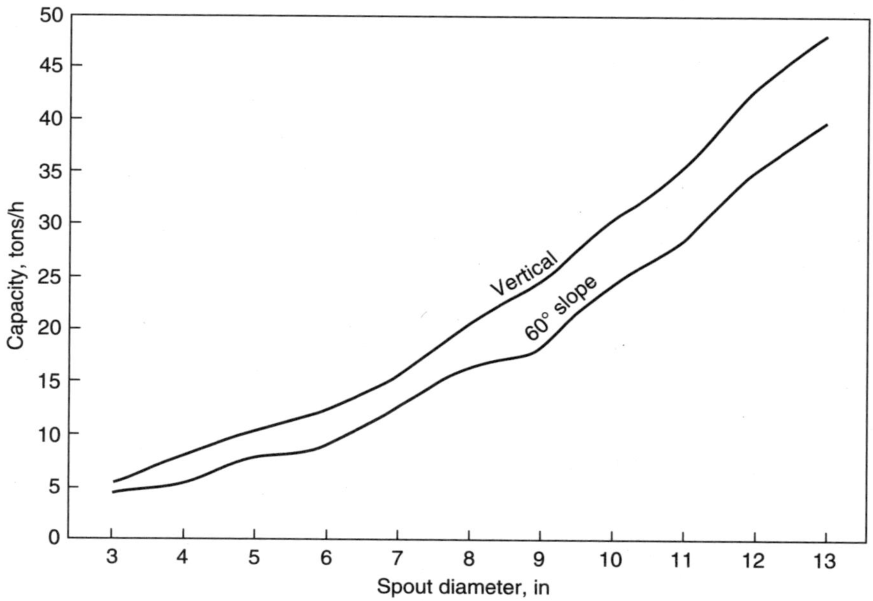

Figure 6.7 Gravity spouting flow capacity. [Notes: (1) powder density 25 to 40 lb/ft^3. (2) Free-flowing granular and aerated powder can be 4 times capacity.]

the capacity or flow rate of the powder and air mixture reqired. For estimating, a powder of 65 lb/ft^3 in a layer 2 in deep would require about 10 ft^3 of air per square foot of media. For large particle powders of low density, say 20 lb/ft^3, it may take 60 ft^3 of air per square foot of media. Manufacturers will provide performance testing to determine the suitability and design parameters.[1]

6.9 Gravity Spout Capacity—Powder

The wide use of gravity flow spouts of various types has usually been a haphazard or quesstimated sizing application with some judgment from past experience. Experienced estimates of flow capacity have now become available (see Fig. 6.7).

Unless experience or tests indicate otherwise, it is recommended that the chart values in Fig. 6.7 be used for sizing either vertical or 60° slope spouts. Lower slopes are not recommended without proof of flowability. However, note that free-flowing granular and aerated powders can flow at rates up to 4 times the charted capacities.

Reference

1. *Pneumatic Handling of Powdered Materials,* EEUA Handbook 15, Engineering Equipment Users Association, London, England, 1963.

Chapter

7

Accessories for Collectors

Every facet of a system needs to help optimize air pollution control, from dust capture to collection efficiency to performance testing and tune-up. The accessories play a big part, especially in controlling the operation and achieving energy reduction and other cost savings. For example, the accessory hardware on a sailboat provides for better control of the sails, allows for better navigation, and provides an all-important indication of the actual boat speed.

Accessory hardware on a dust collector provides for measurement of pressure drop across the filters, allows better control of filter cleaning cycles, helps achieve minimum energy usage and maximum life of equipment, enables the best sequencing and interlocking operation of all the system's equipment, and permits the measurement and recording of the ultimate dust emission levels out of the exhaust. These accessories are vital to good operating control and permit the fine tuning of all facets that achieve optimum operation and collection efficiency.

7.1 Emission Monitors

Three major factors prompt the use of emission monitors for air pollution control. Obviously if the EPA or other environmental control board requires it, this is the first reason to provide and maintain the required monitoring and pollution prevention instruments and records of their measurements. The second reason is to optimize process control, reduce or eliminate air pollution, and minimize costs, all of which can be helped by monitoring emissions. The third reason is self-preservation for each company with air emissions and to help each person responsible be able to sleep well at night knowing that they have controls and alarms to warn of operating failure.

Several decades ago the primary test was a visual sighting of the exhaust stack to see if "smoke" or "dust" was coming out and if it was opaque enough to cause a problem or complaint. A small transparent plastic panel with five levels of opacity was held up in the line of sight toward the stack to serve as a comparison gauge. If the exhaust plume did not exceed 20 percent opacity, it was acceptable. Soon the level was reduced to a maximum of 10 percent, but for 20 to 30 min/day a higher level was allowed for start-up and upset conditions.

Those days are gone now. The requirements for emissions monitoring are very stringent. Currently, the general rules required are covered in the United States by federal law as follows: air emissions are regulated by the Clean Air Act, wastewater is regulated by the Clean Water Act, and hazardous solid waste is regulated by the Resource Conservation and Recovery Act. We will only concern ourselves herein with the Clean Air Act (CAA); Clean Air Act Amendments (CAAA), primarily with particulate dust emissions; and most recently with PM-10 requirements, which cover particulate emissions in sizes under 10 μ.

The CAAA has special provisions for dust particulates smaller than 10 μ in diameter and for air toxins. The regulations are getting even tighter, so both the function and the operating control of collection performance must improve to keep pace.

Standard sampling, testing, monitoring, and reporting methods for stack gases are listed in the EPA specifications, which come from the Code of Federal Regulations, #40 CFR, Part 60, Appendix A, "Selected Reference Methods." Method 5 is often required for sampling determination of particulate emissions from stationary sources, or an appropriate continuous emissions monitor (CEM) may be required.

To meet very low emission levels, and for adequate air toxin control, better techniques and combinations of methods and equipment may be necessary, for example:

1. Improved filter media, membranes, and better filter cleaning methods
2. Initial and operating application of filter aids or dust cakes
3. Leak testing, improved sealing, and better seams on filter elements
4. Improved dust capture or pickup at the source
5. Continuous emission monitors
6. Performance testing and tune-up for continuous top results

7. Secondary and tertiary stages of filtration when needed
8. Accessories to improve the function, measurement, and control of collection performance

New development

Software-based CEM is a model of the system, or simulation program, that uses sensor data from operating units to predict emission levels based on initial hardware emissions measurements. The software model is formulated from available process measurements and from an earlier emissions level sampling that permits the generation of a model of a system's ongoing performance.

Sensor input information from the process is used to update the program's calculated emissions level. It is intended that the software will identify these levels and causes of emissions and then calculate operating set point adjustments to reduce or control them.

The Texas Air Control Board voted in August, 1993, to adopt software-based predictive emissions monitoring systems (PEMS) such as described above. Under the new regulation, companies will be allowed to use PEMS instead of hardware-based systems to comply with the Federal CAAA of 1990. It was announced that efforts are under way to obtain federal approval for nationwide use of PEMS-type systems. It is claimed that chemical process industries could reduce the cost of compliance by more than 50 percent using PEMS.

While there is great potential here, it would seem wise to protect yourself with regular performance testing and a hardware-type CEMS, at least until PEMS becomes a well-proven method. The critical question is not whether the system can simulate performance, but whether it will be equipped to identify all sources of pollution and operating problems. Sensor technology has improved greatly, but what will be the cost of sensing every critical function in a collector?

The first blind flight was made by Jimmy Doolittle around 1930. He used very few instruments, which were relatively crude. We owe him and others a big debt, but 60 years later we still have occasional questions about 100 percent reliability of even our latest digital instrumentation. With all the newer instruments for blind flying, it is still reassuring to break out of the clouds and see that you are over the runway, and at the right airfield. Regular performance testing and a simple emissions monitor, which themselves are not foolproof, are always good operating precautions and cost relatively little. So watch for progress, but be sure to back up your emissions measurements dependably! All sources of emission would have to be detected by sensors, or predicted by PEMS, to be fully dependable and comprehensive.

Sampling train

In EPA Method 5 a sampling train is used to take stack gas samples at specified time intervals to measure the amount of dust particles emitted into the airstream. But organic and volatile organics must also be analyzed if present in the EPA Method 5 sampling train. Volatile organic compounds (VOCs) have become a big concern in air pollution reduction as more has been learned about their effect on the environment. As an example, there have been some major changes in paint ingredients to eliminate the VOCs or to meet the allowable levels.

Continuous emission monitors

These monitors are required if intermittent sampling is not satisfactory and continuous measurement and recording of varying levels are called for at emission sources. The CEMs can be installed directly in the exhaust stack or there can be a sensor in the stack with the control unit located remotely.

The EPA specifications for CEMs are covered in the Code of Federal Regulations, #40 CFR, Part 60, Appendix B.[2] These specifications are applicable to most industrial dust collectors and industrial furnaces and boilers (IFBs). The CEMs can be used for measuring emission levels of dust and/or gases such as sulfur dioxide, nitrogen oxides, oxygen, carbon monoxide, and carbon dioxide (SO_2, NO_x, O, CO, and CO_2).

For dust particulate emissions only, the two major types of CEMs used on the exhaust stack or duct are the light transmission detector of opacity levels and the triboelectric effect sensor of particles impacting the sensor probe.

Opacity monitors

Opacity monitor use began as simple light transmitters and receivers that sensed the level of transmitted light received after passing through an exhaust airstream. Many light detection or extinction principle opacity monitors now incorporate a feedback circuit that permits compensation for long-term variations in the light source intensity, detector response, and window fouling. This allows the opacity monitor to indicate short-term signal changes that measure the dust level in the airstream, rather than indicating long-term fouling effects. Response set points can be adjusted to initiate an alarm or control action based upon the change in light level which indicates excessive dust loading in the airstream.

A light beam, or a collimated light beam, is projected across the diameter or width of an exhaust stack or outlet duct. Dust particles reduce or block the amount of light transmitted through the airstream to a detector, which measures the light level received. The normal light level that is present without any dust in the airstream can be accepted as a base level or zeroed out by adjusting sensitivity set points in a calibrated system. In normal operation then, the measurement of light received indicates the opacity level at the current collection efficiency.

When a filter failure, a broken bag, wear, seam failure, or dust penetration through the filter media permits an increased level of dust particles to be exhausted, the monitor should sense the increased opacity. An adjustable time delay is usually included in the monitor control circuit to allow momentary cleaning pulse releases of slightly higher dust levels, or puffs, to be ignored for from 0.5 s to the maximum cycle time.

The measured signal can be indicated on a meter and is usually also connected to actuate a relay that controls an alarm or, at a higher level of dust loading, to actuate a relay that controls the fan or process equipment. The monitor signal can be, and is increasingly, recorded for either EPA or operating records of dust emission levels. (See Fig. 7.1.) This record can now be made on a computer disk, in which form it is acceptable to the applicable pollution control agency and serves as a backup to chart records.

Each cleaning cycle can usually be identified on the strip chart as a

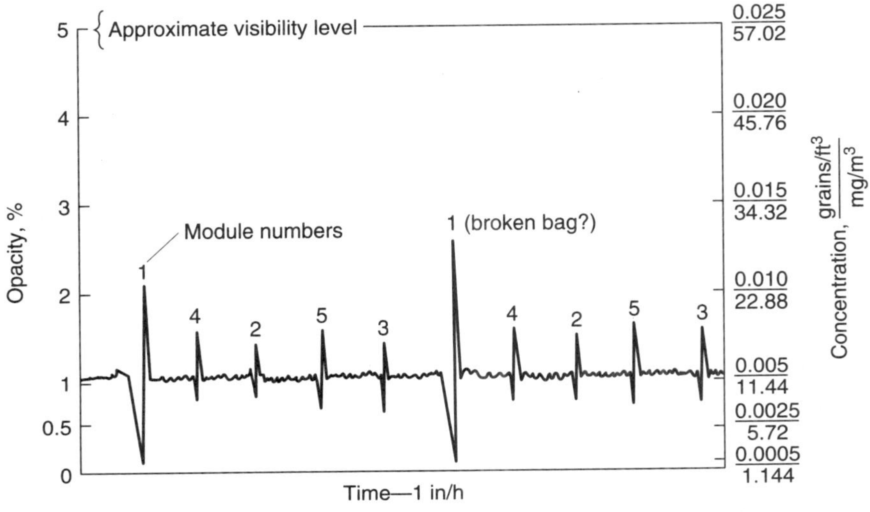

Figure 7.1 Strip chart record—low-level emissions.

momentary pulse of opacity (see peaks 1 to 5 in Fig. 7.1). In steady-state operation of cleaning pulses the levels of opacity increase. The trend or amount of increase can be related to the specific collector module or row of filters. If the increase is gradual, then maintenance may correct the problem at a convenient time or changes in cleaning effectiveness may overcome the increase in dust levels. If the emissions increase rapidly, then an alarm may alert personnel for corrective action and/or a shutdown procedure can be initiated before undue emissions occur.

The emissions detected can be as low as 0.1 percent opacity, or about 0.0005 gr/ft^3 of dust particle loading, on current monitors. The emission level can be correlated to indicate relative levels of toxic or hazardous materials, based on dust content sampling. For a schematic diagram of a typical opacity monitor see Fig. 7.2 (plus a triboelectric monitor and a Pitot tube velocity or flow meter).

Opacity can be calculated as follows, where light transmitted is accepted as 100 percent and light received is the level as reduced by

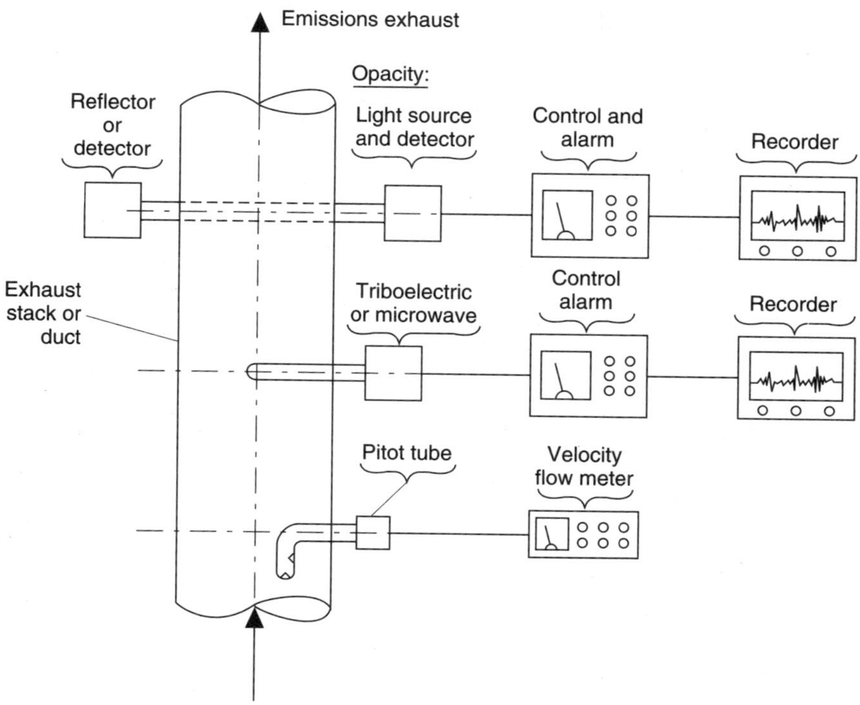

Figure 7.2 Opacity and particulate emissions monitors, with Pitot tube velocity measurement in exhaust stack/duct.

the opacity of the dusty airstream which the light beam must pass through.

$$\% \text{ Opacity} = \frac{(1 - \text{light received})}{(\text{light transmitted})} \times 100 \qquad (7.1)$$

For instance if the light received through the dusty airstream is 98 percent of that transmitted, in whatever units, then

$$\% \text{ Opacity} = \frac{(1 - 0.98)}{(1.0)} \times 100$$

$$= 0.02 \times 100$$

$$= 2.0\%$$

The mass flow of dust emitted, or dust loading in the airstream, is related to the measured opacity as follows:

$$\text{Mass flow (or dust loading)} = W = K_1 \times \frac{P}{L} \log (1 - \text{Op}) \qquad (7.2)$$

where W = mass flow, g/m^3
= dust loading, g/ft^3
K_1 = plume parameter, m^3/m^2
P = particle density, g/m^3
L = duct diameter, m
Op = % opacity

For an incinerator operation with a K_1 of 0.33 to 0.5 (average of 0.415), the average measured opacity of 10 percent produced a mass flow of 0.05 gr/ft^3 (standard) or 114.4 mg/m^3, or 0.1144 g/m^3. Then mass flow is calculated as follows:

$$W = 0.415 \times \frac{P}{L} \log (1 - 0.1)$$

$$= 0.415 \times \frac{5.2328}{2} \times 0.1053605$$

$$= 0.415 \times 2.61638 \times 0.1053605$$

$$= 0.1144 \text{ g/m}^3 \qquad (7.2)$$

where P = 5.2328 g/m^3
L = duct diameter, m

Combination monitoring systems

Combination monitoring systems are available that monitor dust particulate loading or opacity, as well as combustion exhaust gases NO_x, SO_2, CO, and CO_2. Multipoint systems switch to multiple stack sensors on a sequentially timed basis. They are also available with recorders and remote signal outputs. These opacity monitors are available to meet EPA specifications and can be certified for compliance with EPA measurement levels.

Triboelectric detector

This is the second major type of emissions monitor. It measures the small electric charge (relative to the particulate loading) created by the friction of dust particles impacting on a metal probe. The electric charge or signal level is compared to the normal or base signal created by the clean-air flow activity. Any significant increase in signal is measured and can be used to indicate the dust level and/or to activate a relay for alarm or control action when the signal exceeds the adjustable set point. A delay timer can also be used to overlook any normal brief puffs of dust in cleaning cycles.

Current triboelectric CEMs have a dust loading sensitivity down to 0.01 gr/ft^3 (actual) (22.881 mg/m^3) that can be guaranteed. Some are now capable of detecting dust particulate concentrations down to 0.001 gr/ft^3 (actual), or 2.281 mg/m^3. It is understood that the EPA is

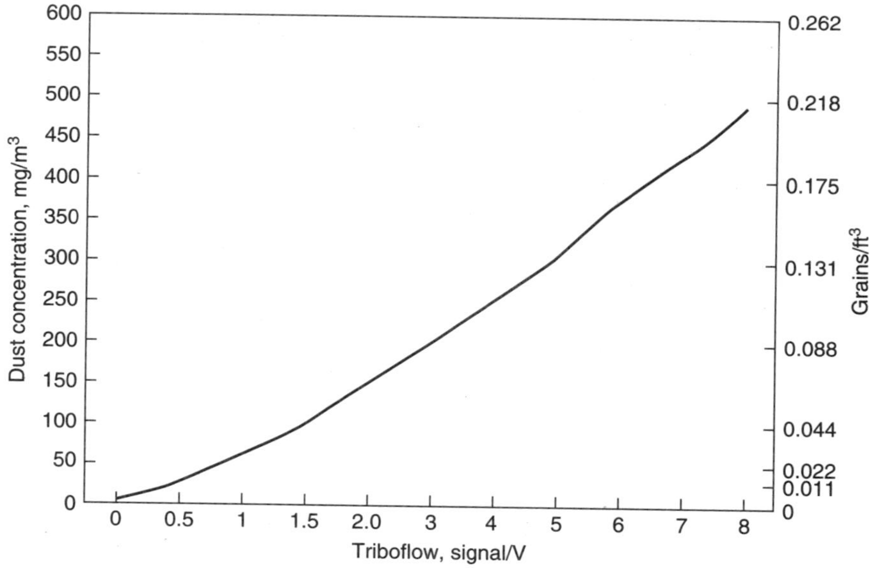

Figure 7.3 Triboelectric monitor—emissions signal versus dust load.

considering the certification of these triboelectric CEMs for measurement or recording of dust emission levels. Their application to industrial dust collectors offers the option of a theoretically more dependable or trouble-free operation than the opacity monitor, because of their lower operating maintenance and calibration requirements (see Fig. 7.3).

Up until a few years ago the generally accepted and utilized emissions monitor was the opacity or visual type. Varying opacity readings can occur due to dirty glass ports in the duct and any other light transmission decreases, which necessitate air purge blowers and adjustment of sensitivity. While automatic sensitivity adjustment circuits and an air purge have improved reliability, their added complexity has given impetus to the alternative use of triboelectric monitors which do not have this drawback.

As air pollution regulations and emission limits have become more stringent, the allowable dust loading level emitted has been reduced. Current limits are often down to 0.01 gr/ft^3 (standard), and even lower levels down to 0.005 gr/ft^3 (standard) are being set. As better pollution control is required to improve environmental air quality, both the measurement accuracy and dust collection efficiency must be improved. Some solutions to improving collection efficiency are as follows:

1. Filtration media with maximum capture efficiency, such as porometric membranes, dual density fabrics, electrostatic augmentation, or cartridge filters with membranes and wider pleat spacing
2. Lower filtration velocity through the media, which requires a larger filter area for the same airstream volume
3. Better inlet design for lower air velocity and for more uniform distribution of airflow to the filters
4. Evaluate the use of filter aid powders or the controlled buildup of process dust cake.
5. Lower can or passage velocity in the filtering zone to reduce particulate entrainment, and thus preserve as low a filtration velocity as possible by maintaining the full open area of the filters

CEM costs

Continuous emissions monitors for a large IFB stack can run from \$15,000 to \$25,000, or more. Cost for CEMs of either opacity or triboelectric detection type for most industrial plant stacks and collector exhausts is around \$6000 to \$8000. Multipoint monitors for up to six emission source sensors per monitor will cost from \$1500 to \$2000 per sensor measurement point.

Opacity monitors reportedly have a sensitivity from 2.0 percent opacity = 0.01 gr/ft^3 (standard) = 22.881 mg/m^3, down to 0.2 percent opacity = 0.001 gr/ft^3 (standard) = 2.288 mg/m^3. Triboelectric monitors reportedly have a sensitivity down to 0.1 percent opacity = 0.0005 gr/ft^3 (standard) = 1.144 mg/m^3. The higher sensitivity opacity monitors as used on IFBs also are reported to reach the 0.1 percent opacity range but at costs of around $25,000 each.

Costs for triboelectric single-point "broken bag detectors" have come down to about $1500 to $1750 each. These will alarm and/or provide relay contact closure for control action on dust emissions in the 0.01 gr/ft^3 (standard) = 22.881 mg/m^3 = 2.0 percent opacity, down to 0.005 gr/ft^3 (standard) = 11.44 mg/m^3 = 1.0 percent opacity range.

This cost level would probably justify a broken-bag-detector emission monitor on nearly every dust collector of any significant size whether required or not, especially as a safeguard and/or for record keeping and liability protection. Their reliability has been greatly improved, and for this entry-level monitor the application and maintenance is very simple. As a recent *Chemical Engineering* article said, "If data aren't required now, they soon will be."

Changes and developments in emission control

Information on the CAAA is now accessible by computer or phone on latest changes in emission regulations. The electronic bulletin board operated by the EPA's Office of Air and Radiation lists signed regulations issued since March 1992. Access is possible by most personal computers by dialing 1-919-541-5742. Once connected, go to the CAAA menu item. If you have difficulty you can call the Technology Transfer Network at 1-919-541-5384.[2]

Air toxins can be controlled in many ways depending upon the nature of the toxin. Solid and liquid waste may be burned or vitrified in a cement, asphalt, or other kiln. Cement kilns typically operating around 2450°F (1343°C) will destroy organic compounds and vitrify inorganics and heavy metals in the inert cement clinker output.

Ammonia injection, selective catalytic reduction (SCR) or selective noncatalytic removal (SNCR) can be used for NO_x removal in combustion flue gas usually after particle filtration. Catalysts can remove VOCs from process exhausts. Now microorganisms can oxidize VOCs to achieve 90 to 99 percent removal and produce CO_2 and water byproducts.

Lime injection and sodium bicarbonate sorbents for SO_2 emission reduction, NO_x removal discussed above, and filtration for particulate control are being used to achieve acceptable levels of emission. A combination filter collector has been developed to remove or control SO_2, dust particulate, and NO_x in a single enclosure, the Babcock and Wilcox patented SO_x-NO_x-Rox-Box, SNRB process.[3]

Mexico is now reportedly trying to accomplish in 5 years of air pollution control what it has taken the United States to do in 20. As other countries shoulder their share of improving the world's environmental control, industry in the United States may begin to achieve a better competitive position.

7.2 Level Sensors

How do you know what you've got if you can't see it, feel it, or hear it? Level sensors provide a measure of the quantity or fill level of bulk solid or dust contents in hoppers, and they help detect operating problems. The cost can vary from about $100 per point up to $5000 or more for continuous level and loss-in-weight detectors and controls. Accuracy is usually more than adequate for dust collection level sensing service.

Paddle, diaphragm, and tilt switch

These are some of the first level-sensor types used and are still valuable, dependable, and economical. They provide a single-level measure only, so two would be needed for both high- and low-level sensing, for instance. The paddle and the simple tilt switch single-point level detectors would project into or be located within the hopper or collector housing. The diaphragm design usually mounts on the housing surface with a 6- to 8-in diameter hole cut out to expose the diaphragm to the bulk solid or dust material. One advantage of the diaphragm type, other than its simplicity, is its easy enclosure for higher temperature or explosive atmosphere service at low cost. The cost for these simple single-point level sensors can range from around $100 to as much as $600 each.

Radio frequency, ultrasonic, and capacitance

These later developments provide more options such as remote indication, analog signal output, PLC and computer input, higher-temperature service, self-checking, and detection of light density material down to 2 lb/ft^3. Their cost can run from $75 for the lowest-cost capacitance type to several thousand dollars for a continuous level reading radio frequency type.

Continuous level

When the level needs to be read over a range from low to high and that measurement needs to be indicated or transmitted to a remote indicator or printer, or for control action, some form of continuous measurement is useful. This can be the simple yo-yo type in which a motor drive raises and lowers a weight hung by a cable from a reel and now can provide output signal data. Load cell or strain gauge sensors provide a weight measurement of contents with a weighted cable or with strain gauges on the support structure for a continuous level reading and data output. Ultrasonic or radio frequency sensors also provide continuous measures of the distance to the dust surface with various data outputs. Another type available is a radiation source that detects continuous level changes by an ion chamber or scintillation counter and transmits readings to a remote computer. Costs can vary from around $500 for the ultrasonic or yo-yo type with some optional control adders, up to around $4000 for the continuous weight or radiation-type continuous level sensors.

7.3 Fire and Explosion Venting, Suppression, and Extinguishing

Wave a red flag in your mind early and often about these safety concerns! The best safety measure is to be sure to learn all there is to know about the hazards of the particular dust being handled. The basic prevention controls are good housekeeping, grounding, and safe dust cleanup procedures to prevent sparks, fires, and explosions. Dust accumulation should be cleaned up regularly or prevented to reduce the risk. Electrical sparks from wiring faults or electrostatic buildup and discharge need to be prevented by proper grounding, good wiring, and isolation by protective enclosures, such as NEMA standards type #7 and #9.

Then there are various venting, suppression, and extinguishing methods. Whatever the product or process, wave that red flag in your mind about these hazards during project design, and be careful to consider taking adequate protective measures!

Pressure containment—passive protection

The containment of explosions in a vessel is an older method and is often expensive and dangerous if the explosion pressure exceeds the safe design pressure, since the vessel then becomes a bomb. NFPA standard #69 provides guidelines for calculation of the necessary design strength. Generally, existing or relatively large vessels will not benefit from containment design. However, for new or relatively small process vessels it may be advantageous.

A variation of this is the directed containment that was practiced as much as 200 years ago in the manufacture of explosive dynamite and gun powder. Three walls were stoutly built and the fourth wall or the roof was nonexistent or purposefully designed to be weak. The weak wall or roof was directed toward a river or other "safe" direction so a blast would be less harmful.

Hagley Museum in Wilmington, Delaware, preserves some old DuPont powder mill buildings of this design from the early 1800s. They are not in vogue now because of the potentially high damage liability and the fact that some employees might be within the process building when it blows. Other examples of the containment method of protection is the test of atomic bombs in underground chambers, or the centuries old design of firearms where the explosive force is directed out the barrel opening. This is also the basis of rocket engine propulsion, where some of the forces of combustion are released out of the exhaust pipe, thus allowing the rest of the force to exert an unbalanced propulsion thrust on the forward wall of the combustion chamber.

Explosion venting

This is now a common and reliable means of control, but it can be relatively expensive. However, it is certainly inexpensive compared to the alternative damage with no protection. Similar in principle to the weak wall or to a rupture disk, vent panels are designed to protect a process vessel, such as a dust collector, by failing at a set or predesigned pressure level.

Explosion venting membranes, panels, or fasteners are designed according to the following code guidelines in the design of deflagration venting systems:

1. ANSI/NFPA standard #68 on explosion venting
2. BOCA—The National Building Code
3. ICBO—The Uniform Building Code
4. SBCCI—The Standard Building Code

Guidelines for explosion prevention systems are covered in ANSI/NFPA standard #68. Again, the insurance company responsible at the installation site may have its own guidelines which need to be met in addition.

Equations and nomographs are presented in NFPA standard #68 for estimating the safe vent area needed based on the hazard rating of the dust, the volume of the enclosure to be protected, and the estimated pressure rating or design strength of the enclosure vessel. A brief list of explosive hazard ratings for dust and powder materials is presented therein.

TABLE 7.1 Explosive Hazard Ratings for Typical Dusts

Dust material	K_{st}, bar · m/s
Lowest Class (K_{st} less than 100 bar · m/s)	
Adipic acid	97
Bronze	31
Calcium acetate (92-μ median)	9
Calcium acetate (85-μ median)	21
Cellulose, pulp	62
Charcoal, activated	44
Charcoal, wood	10
Coke, petroleum	47
Corn	75
Egg white	38
Lactose	81
Melamine, molded (wood flour and mineral-filled phenol-formaldehyde)	41
Milk, powdered	28
Peat, 22% H_2O	69
Polyacrylamide	12
Polyvinyl chloride	46
Polyvinyl chloride, ethylene/vinyl acetylene emulsion copolymer	95
Polyvinyl chloride, ethylene/vinyl acetylene emulsion suspension	98
Soot, pine	26
Sugar, beet	59
Sugar, milk	82
Tapioca	62
Medium Class (K_{st} between 101 and 200 bar · m/s)	
Ascorbic acid	111
Calcium stearate	132
Carboxy-methyl-cellulose	136
Coal, bituminous	129
Dextrin	106
Epoxy resin	129
Iron carbonyl	111
Lampblack	121
Lead stearate	152
Lignite	151
Melamine, molded (phenol-cellulose)	127
Melamine resin	110
Methyl-cellulose	134
Milk, nonfat, dry	125
Paraformaldehyde	178
Peat, 15% H_2O	157
Phenolic resin	129
Polyacrylonitrile	121
Polyethylene, low pressure	156
Polypropylene	101
Polyvinyl acetate/ethylene copolymer	119

TABLE 7.1 **Explosive Hazard Ratings for Typical Dusts** (*Continued*)

Dust material	K_{st}, bar · m/s
Medium Class (K_{st} between 101 and 200 bar · m/s)	
Polyvinyl alcohol	128
Polyvinyl butyral	147
Sodium ascorbate	119
Sodium stearate	123
Soy flour	110
Starch, rice	101
Starch, wheat	115
Sugar	138
Sulfur	151
Terpene-phenol resin	143
Urea-formaldehyde/cellulose, molded	136
Whey	140
Zinc (10-μ median)	125
Zinc (below 10 μ)	176
Highest Class (K_{st} over 200 bar · m/s)	
Aluminum	415
Anthraquinone	364
Cellulose	229
Cork	202
Magnesium	508
Polymethyl acrylate	269
Polymethyl acrylate, emulsion polymer	202
Starch, corn	202
Wood flour	205

Lacking a specific reference, an actual test of the dust in question would be required for safe explosion vent design. One manufacturer's list of "hazardous materials combustion classification" breaks the materials into three classes, based on their NFPA-determined K_{st} or deflagration index, as shown in Table 7.1. The K_{st} is equivalent to the maximum dP/dt in bar · m/s.

A typical nomograph from NFPA standard #68 (see Fig. 7.4), will provide a preliminary estimate of required vent area. The first step in using the nomograph is to determine the K_{st} of the material from the above or the smaller NFPA list, or have a test run to determine the explosiveness or the deflagration index K_{st} of the specific dust to be collected. Then the volume and the safe design strength of the collector vessel must be determined and these figures set in metric units. The lowest static design vent release pressures, P_{stat}, are normally applicable to dust collector applications at 0.1 bar gauge.

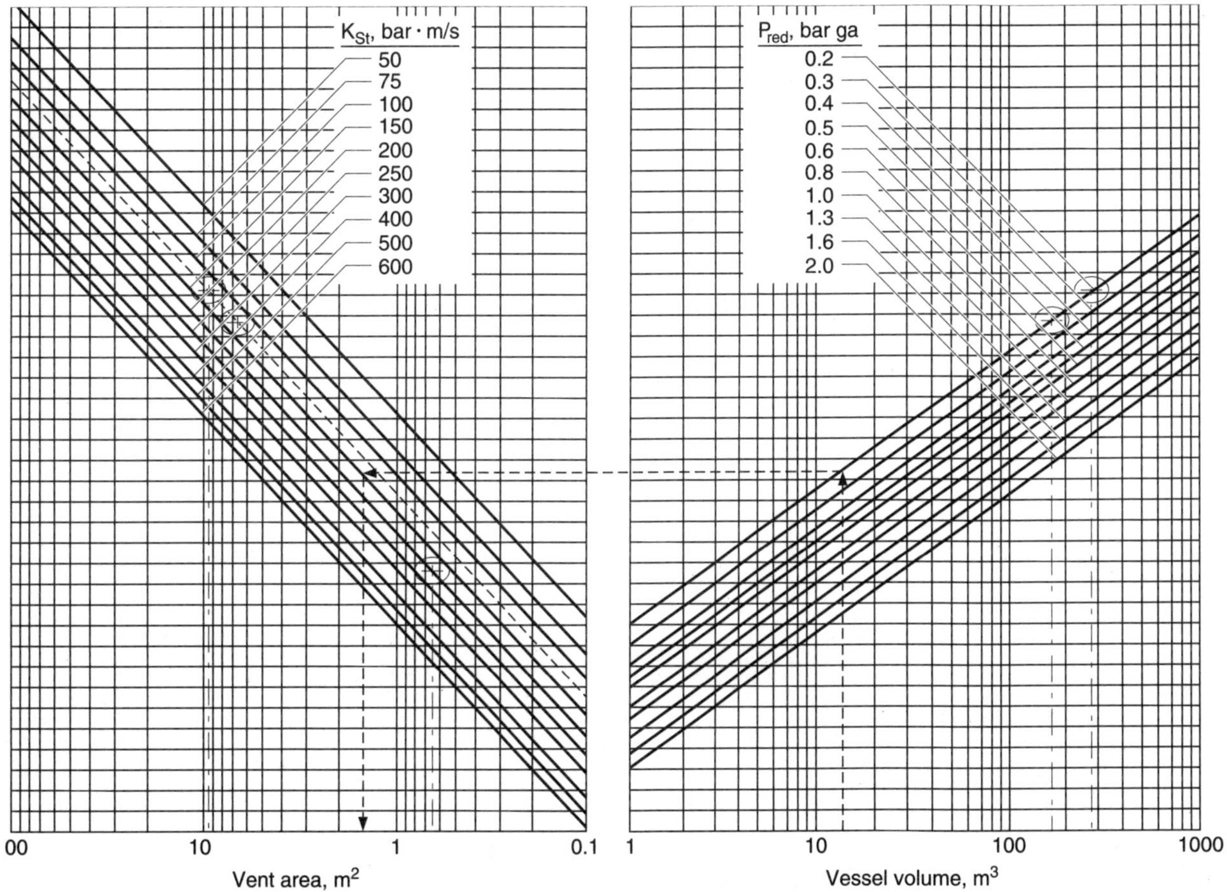

Figure 7.4 NFPA #68 venting nomograph for dusts. (P_{stat} = 0.1 bar gauge = 1.44 lb/in^2 gauge.)

Then using the NFPA nomograph for the lowest P_{stat}, enter the nomograph at the vessel volume in cubic meters Go up vertically to the intersection with the appropriate line for P_{red}, or maximum pressure allowable or safe in the vessel during venting. From this intersection, proceed horizontally across to the intersection with the appropriate K_{st}, or class of dust explosiveness. From this second intersection drop down vertically to the required vent area in square meters for the application.

This calculation can be performed on a computer by some suppliers now and would usually be more accurate than the nomograph. However, remember that the final accuracy involves the strength and volume calculations, safety factors chosen, the effectiveness of fabrication on the strength of the collector vessel, changes in operating and maintenance conditions over the life of the system, and many other factors that will affect the overall strength and safety of the system. So, use all the available accuracy possible, but don't overlook the major factors. As Dr. James Kennedy says, "Don't sweat the small stuff!", at least not to the exclusion of the larger factors and the overall safety and effectiveness.

Another method for determining the vent area required, and probably the most commonly used, is the vent ratio relative to the combustion class, explosion hazard, or deflagration index. This is the vent ratio R, of vent area V needed for the volume C of the vessel, collector, or building involved, as shown in the following equation.

$$\text{Vent area } V = \frac{\text{volume } C}{\text{vent ratio } R} \tag{7.3}$$

The nominal vent ratio is taken from Table 7.2.

TABLE 7.2 Hazardous Materials Combustion Classification for Explosion Venting

Hazardous material combustion class	K_{st}, bar · m/s	Vent ratio R, ft³/ft²
Broad classification range		
Low	Up to 100	80
Medium	101–200	50
High	Above 200	30
Usable vent ratio levels		
Low	Up to 100	80
Medium	100–125	64
	126–150	58
	150–200	50
High	201–250	44
	251–300	39
	301–400	34
	Above 400	30

Then for a collector vessel with dimensions of 100 ft^2 by 25 ft high, and thus a volume of 2500 ft^3, for a medium class material of epoxy resin with a K_{st} of 129, the broad range vent ratio R is 50 from Table 7.2. Rearranging Eq. (7.3),

$$V = \frac{C}{R}$$

$$= \frac{2500 \text{ ft}^3}{50 \text{ ft}^3/\text{ft}^2}$$

$$= 50 \text{ ft}^2$$

But from Table 7.2, the usable vent ratio R is 58, so from Eq. (7.3)

$$V = \frac{C}{R}$$

$$= \frac{2500 \text{ ft}^3}{58 \text{ ft}^3/\text{ft}^2}$$

$$= 43.1 \text{ ft}^2$$

So the 50-ft^2 vent area requirement would necessitate a 5- × 10-ft vent for a single opening, a somewhat unwieldy vent panel or door. This is also the case for the 43-ft^2 vent area requirement. By using four or five separate vents, a better or more workable size for installation, access, or ducting to the outside would result. The 50-ft^2 vent area requirement could be met by using five 2- × 5-ft (24- × 60-in) vents that would usually be better suited for installation. An alternative would be five 20- × 72-in panels that allow an easy fit between reinforcing ribs at 24-in centers and make a good access door size. Six vent panels would accomplish some quantity price reduction, as well as allow for one to be tested. Seven would add a ready spare for future replacement.

The 43-ft^2 vent area requirement could be met by using four 11-ft^2 panels of 20- × 79.2-in size, or five 20- × 62-in size panels. In any event the 43-ft^2 total saves about 7 × \$135/ft^2 = \$945.00. This is well worth consideration in the long run for the savings at no loss in safety, as long as the insurance carrier approves.

A commonly used vent ratio R in the industry has been 35:1 as an all around estimating guide. By reference to the classifications in Table 7.2, 35:1 would be a conservative ratio for all but the highest levels of K_{st}. In fact as a correlation to the NFPA equations and nomograph results shown in Fig. 7.4, a gradual decrease in vent ratio would be required as K_{st} gradually increases up to the 600 maximum with a V of 30. The nominal 35:1 ratio would be needed only for a K_{st} that exceeds 375.

Skylight-type explosion relief vents can be mounted on the roof or wall of a collector to provide daylighting, and they are available in larger sizes from 4 × 4 up to 6 × 12 ft. They may be rated to open at 10 lb/ft^2 or 2 in WG minimum. They come with a safety cage designed to withstand a 700-lb impact to make it safe for workers. And they can be designed for the expected wind loads.

This division into several smaller or standard size vents may be advantageous for several reasons. First, one large vent may be unwieldy in design, location on the vessel, cost, and venting to the outside. Second, it may be less expensive to purchase several smaller vents, especially for quantities of three or more vents, and where the certification requires testing of one vent to failure by the manufacturer. Third, there is the possibility of standardizing on several vent sizes in a plant, so that the initial cost may be reduced by annual purchases of larger quantities of these standard sizes. Fourth, the cost and time for obtaining spares for replacement when needed are both reduced. Replacement may be needed for either an actual explosion occurrence or when the panels are judged to have aged out. Most vent membranes, foam panels, or fasteners will need to be replaced on a regular schedule of from 2 to 3 years, depending upon exposure conditions. Spring-release latches or magnetic catches should be tested and reset if necessary on a 1- to 2-year basis to ensure that release for venting will still occur at the design pressure.

The best course of action in any case is to get recommendations and approval from both the insurance carrier for the proposed collector installation and the explosion vent supplier, along with written assurance of safe design from the supplier. Vent suppliers can perform tests for a fee on several sizes of test vessels, and they can be conducted following both the NFPA and the International Standards Organization (ISO), formats to ensure that all tests are performed as required and data are referenced and pertinent.

Vents are intended to prevent the destruction of an enclosure due to an explosion that exceeds the strength of the vessel, caused by a dust or vapor explosion or by a mechanical or electrical system failure. Relief of the explosive pressure is possible by the use of low-mass, accurately designed and predictable release pressure venting devices. The NFPA standard #68 calls for vent panels to weigh no more than 2.5 lb/ft^2 to ensure ready opening and to allow safe restraint of the panel in an explosion.

All vents must be designed to withstand the negative operating pressure as well as the positive pulse pressure developed in collector cleaning cycles. So some of the unsupported membranes or panels will not withstand the pulse pressure cycling, but most vents are available with adequate strength or support for both the negative operating pressure and for pulse-jet cleaning pressures.

Vent designs in use range from plastic foam sheets; hinged panels; rupture membrane panels of plastic, stainless steel, or aluminum; to fasteners or spring-loaded latches that release a panel or door at a calculated pressure. Thus the calculated release or failure pressure will safely vent an enclosure. Engineered and test-certified panels and fasteners are calculated to fail or vent at a safe pressure rise just below the safe design pressure or vacuum level (see Fig. 7.5).

An unvented enclosure would be exposed to the resultant maximum pressure P_{max} reached in an explosion. This may be in excess of the safe design pressure P_{des} of the vessel. If the vessel, or collector enclosure, is properly vented, the resulting reduced pressure P_{red} at vent release would be less than the P_{des} and should not cause any damage (see Fig. 7.6).

While venting is a desirable means of controlling explosive pressures, the operation and overall design of the vents must be considered. By-products of an explosion are released by the vent and must be directed safely, usually outdoors, where no further damage will occur. Limit switches can be used to detect and signal a vent panel or door failure, if you have not already heard a report!

However, some other explosion concerns are

- Relieving the pressure does not extinguish a flame if present, so sprinklers, inert suppression, or other extinguishers may be needed.
- Venting does not stop flame propagation to adjacent areas or equipment, so some type of isolation valve may be needed.
- Blast energy must be directed away from personnel or other hazardous locations, such as through a short duct to a safe outdoor area. This outlet duct should be straight and not more than 3 m long to minimize back pressure. There should not be more than a 5 percent reduction in cross-sectional area due to any restriction in the duct.
- Replacement vent panels or fasteners should be readily available or spares kept on hand. A spare panel costs less when bought with the original panels. Some larger users standardize on selected sizes of panels and maintain a stock on hand. That way they make annual purchases in quantity at lower contract prices for these standard sizes.

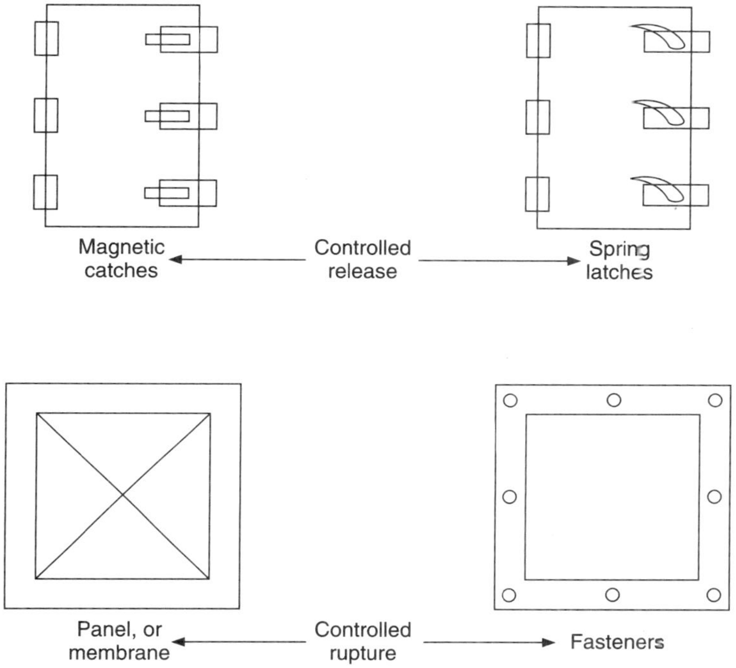

Figure 7.5 Typical explosion vent designs.

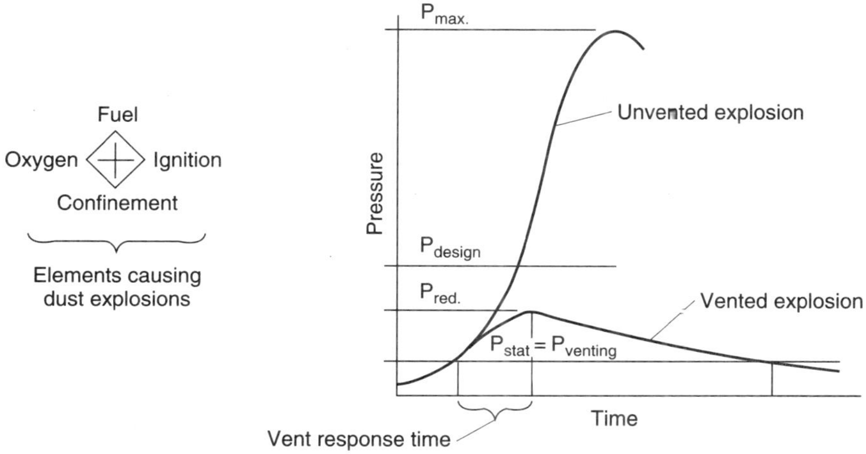

Figure 7.6 Explosive pressure buildup or rise in unvented and vented explosions, with elements causing dust explosions.

- Spring-loaded and magnetic latches provide a reusable feature and also permit use of the panel as an access door. The latches must be set and maintained for the correct release pressure rating.
- Simple screw fasteners with a calculated strength, or a carefully reduced cross-sectional area like a shear pin, have an obvious cost advantage over most panel or membrane designs. They should also be easily and quickly replaced.
- More economical designs are still needed and should help broaden the use of safe explosion vents.

A typical explosion vent specification sheet, or information form, is very helpful in gathering sufficient data for a good evaluation and design selection. It also forces consideration of the process exposure conditions. These operating conditions and the volume of the collector vessel must also be determined for adequate design. (See Fig. 7.7.)

The cost for explosion vents can vary from a few dollars for a "homemade" design using fasteners, latches, or counterweights and hinges, up to around \$135.00 to \$175.00/ft^2 of vent area for a certified manufacturer's panel, membrane, fastener, catch, or latch type design. Many low-cost foam panels have been used on sawdust or grain collectors that cost around \$50.00/ft^2. All these should be approved by the insurance carrier before their installation to meet approved safety and NFPA guidelines, to protect against liability, and to explore possible cost reductions that are acceptable.

Product vent testing by a reputable manufacturer may cost in the range of \$1500 to \$2000. There may be lower-cost tests available, but consider that at some point there needs to be a known explosiveness rating for a product in order to arrive at the optimum protection and cost. Plus the safety and liability protection must be dependable. A typical test will require about 2 liters (L) of representative product.

Inerting

This technique introduces enough inert gas, such as nitrogen or carbon dioxide, into an area or enclosure to reduce the oxygen level below the combustion minimum. This is around 14 to 16 percent for most flammable dusts, but check it out before design or installation. Another possibility is to use an inert gas in the process airstream, and/or in the pulse-jet cleaning system, in place of ambient air. Cost is a factor for inert gas, but it may eliminate the need for other controls.

A. Process conditions
1. Process media, explosive hazard, and percent
2. Vessel design pressure
3. Indoor or outdoor
4. Discharge ducts, size and length
5. Flow velocity
6. Operating pressure and temperature
7. Maximum negative pressure (vacuum)
8. Cyclic/pulsating condition
9. Corrosive environment
10. Wind loading
B. Vent area determination
11. NFPA #68, VD13673, *Chemical Engineering,* other guidelines
12. Description of dust and gas
13. K_{st} or class of dust
14. K_g or fundamental burning velocity (S_u)
15. Process vessel volume and L/D ratio
16. Vent release pressure (P_{stat})
17. Maximum pressure during venting (P_{red})
18. Vessel internal surface area
19. Required vent area
C. Explosion vent selection type
20. Type or designation
21. Will vacuum exceed 50 percent of minimum burst pressure?
22. Will positive exceed 50 percent of minimum burst pressure?
23. Is vent intended for both vacuum and pressure release?
D. Explosion vent specifications
24. Nominal size and number
25. Type release and design
26. Specified burst pressure and tolerance
27. Materials of construction
28. Accessories, mounting, and fastener specifications
29. Limit switch burst detectors or alarms
30. Other requirements, such as outlet ducts

Figure 7.7 Explosion vent specification sheet.

Suppression systems

These have been used since the early 1950s with good results and are the only technique that stops an explosion once it has begun. A dependable system requires a dependable sensor, then a proven extinguishing agent, and a fast-response control system that will always snuff out an explosion before damage can be done. Explosion suppression is a prime alternative to venting where there is not a safe way to vent the overpressure outdoors or if there are toxic or environmentally hazardous materials in the airstream.

Explosion suppression should interrupt the deflagration before destructive pressures are developed. This method also limits the likelihood of a fire following the explosion and reduces the chance of a secondary fire from the flame front that is propagated from many explosions.

Most explosions take something like 200 ms to reach full pressure. The beginning stage is relatively slow and then accelerates rapidly to the full unsuppressed pressure of the explosion. The suppression system must detect and respond during the early stage of this "slow" time rise. There is not much time to respond with any preventive measure, but it is being done!

It takes around 25 ms to detect the initial pressure rise and another 25 ms to respond and emit the suppressant into the vessel. However this is about half the time available before pressure builds up to the vessel's maximum safe design pressure. So there is only about 100 ms available to complete the response and to allow some time for safety margin. The chart in Fig. 7.6 shows the relative rise in pressure over a 200-ms total time period for vented and unvented explosions.

Explosion suppressors may use a 90° elbow as part of the container volume. A 4-in diameter full flow opening is achieved by using a rupture disk as the discharge valve for fast response. When the detector and the controller signal the suppressor system, the fast-response release uses a blasting cap to blow the rupture disk open faster than a solenoid and mechanical valve could respond.

The extinguishing agent can be steam, water, Halon, inert gas, or dry chemicals. A combination of Halon and dry chemicals has been very effective. Steam has been used on wood dust in furniture plants successfully for some years in suppressing flames or sparks detected traveling down a duct prior to reaching the dust collector. If it takes 200 ms to respond with a steam valve with a safety margin, and the airstream velocity is 4500 ft/min (75 ft/s), then that 0.2-s response must have 15-ft minimum length of duct between sensor and steam valve in which to respond.

Isolation and suppression systems

Isolation systems use a similar detection and control system but then either actuate a fast-acting knife gate or inject a chemical suppressant at a safe downstream distance. The aim is to prevent flame propagation through ducts or conveying lines to downstream equipment (see Fig. 7.8).

The isolation valve or suppressant injection point location is critical to the success of an isolation system, like the steam-suppressed system described above. When allowing for response time relative to travel speed, both the airstream and the flame propagation velocity must be considered.

The detector or sensor in either system is vital and must be reliable in

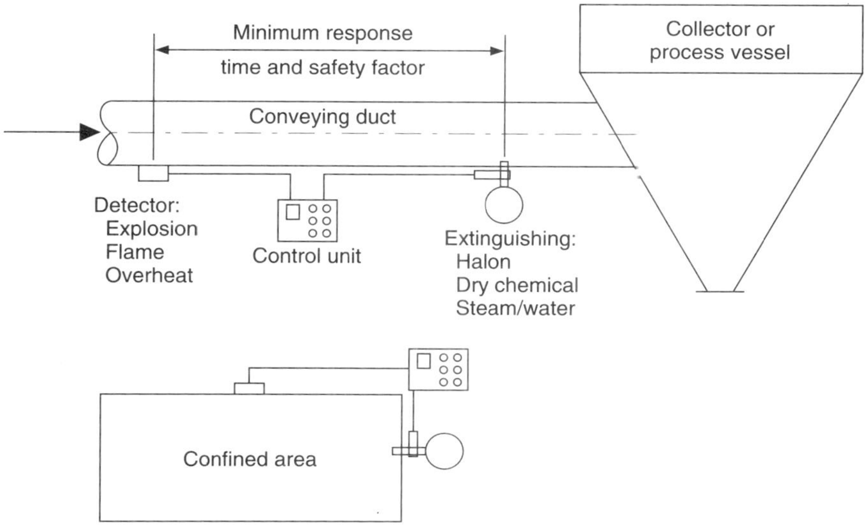

Figure 7.8 Explosion and fire detection and suppression.

the exposure environment. Usually dusty conditions would require a sensitive pressure detector, which can be enclosed so that it is not affected by the dusty atmosphere. Relatively clean atmospheres could use an optical detector, such as an ultraviolet or infrared sensor. The pressure detector can be adjusted in the field and may respond to as little as $\frac{1}{10}$ lb/in^2 gauge. The same order of time response is required, as calculated above, with usually less than 100 ms overall maximum allowable.

Multiple detectors can be used in either system, and multiple isolation valves or injection heads can be included in one control system. Standby or uninterruptible power may be needed in some installations, but most dust collector systems will go down when power fails, so only unusual conditions would require a backup power supply. These extinguishing or suppressing agents must usually guard against reignition after snuffing out the initial explosion. Again whatever method is used should be approved by the insurance carrier and certified by the manufacturer.

Flame arresters

This is a passive safety device that stops the travel of flame in pipelines and ducts from spreading beyond the arrester. They prevent the travel of flame from a flammable group B, C, and D gas or vapor under deflagration and detonation conditions. As one manufacturer says, "There is no standard flame arrester, each is matched to the particular application."

7.4 Inlet Airstream Distributors

Often overlooked before dust collection became scientific, inlet velocity and diffuser designs are now being given the important consideration they deserve. Reduced pressure losses result from a good transition from a conveying duct size and velocity into the collector enclosure at a gradual 15° to 30° flare. Velocity should be reduced to below 2000 ft/min when delivered into the hopper. Then the can velocity should be reduced to the 200-ft/min range before entering the filter passage area as discussed in Chap. 4. In between these two regions the inlet air needs to be distributed evenly across the full cross section of the filter collector or hopper.

Tangential inlet

This has been used on larger-particle-size dusts and chips to separate the large particles by cyclonic action in the hopper entry. It has served well on wood dust and other large-size particles but must have a high inlet velocity for good cyclonic separation. At best it provides a collection efficiency similar to a cyclone, or around 75 to 90 percent. This may serve a valuable function to reduce a heavy dust load to a manageable level for filtration. But bear in mind that maintaining even a limited range of particle sizes presented to the filters can improve the permeability of the dust cake on the filter surface. So consider the overall efficiency and cost of the collection combination and go for the best system (see Figs. 7.9 and 7.10).

Straight or flared transition and inlet

This type can still cause abrasion of filter elements and walls and possibly reentrainment of already collected and cleaned dust at 2000 ft/min (see Figs. 3.7 and 7.10).

Flared and baffled inlet

An angled plate at the end of the inlet transition can help "knock down" some particles and distribute the dust flow somewhat. It is satisfactory on many dust products but not generally on very fine powders.

Flared, baffled, and perforated inlet

Baffle perforation improves distribution of the airstream and dust, but dust can accumulate in the perforated holes, which will cause it to be less effective.

Figure 7.9 Tangential inlet reverse-air filter collector, reverse-air cleaning, and system suction fans visible on top, with vertical exhaust stack.

Flared inlet with diffuser

By providing a series of deflector or diffuser plates, strips, or rings in some fashion across the inlet airstream as it crosses the hopper, a fairly uniform distribution of airflow can be achieved.

- The Cascadair uses a series of successively smaller doughnut-ring plates that divert a portion of the air volume at each ring. Tests have shown a more uniform distribution of airflow to all filter areas, resulting in lower maximum velocities entering the filter passages across the full cross-sectional area.[12]
- The ladder baffle is another diffuser design that uses a series of flat plate strip baffles across the full cross section of the entry to the filter region. This may provide the most uniform air distribution and lowest maximum velocities entering the filter passage area.[13]
- The pelican bill inlet is a wide transition design that brings in airflow to the full width of the hopper at usually lower than 2000 ft/min and distributes the airflow across the full width. This reduces the need for baffles but would still direct more air to the far wall so that distribution may not be as uniform as the baffle designs above.

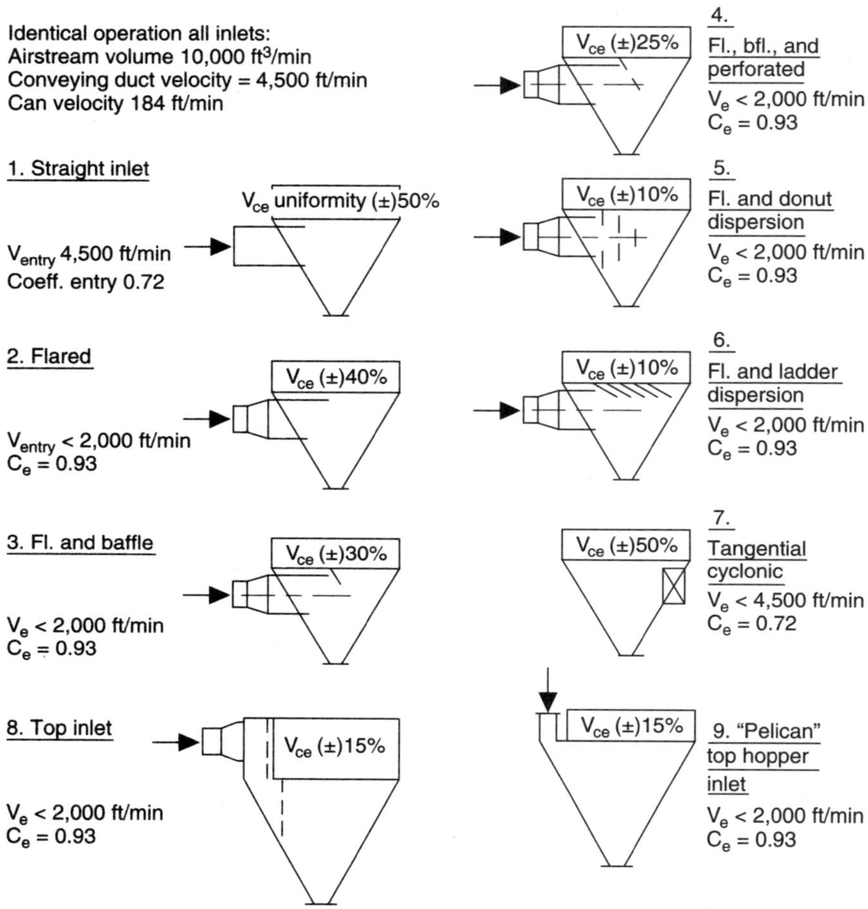

Figure 7.10 Collector inlet designs, with typical operating factors.

- Side or top entry is a refinement of the pelican bill inlet with the addition usually of baffles or expanded metal diffuser sheets across the inlet airflow. In the transition the velocity may be around 1000 ft/min leaving the first expanded metal diffuser, and reduced to around 200 ft/min in the hopper or side wall leaving the second diffuser plate where the air flows into the filter passage area (see Fig. 7.10—5, 6, 8. 9).

These last four inlet diffuser designs are apparently the best in providing uniform distribution and lowest maximum velocity in the filter inlet area. When entering the filter passage area, the can velocity must still be reckoned with by providing adequate passage area. But many advantages would result from using one of these improved inlet designs to provide

- Successful use of higher filtration velocities
- Minimized reentrainment of dust
- Lower pressure drop across the filter elements
- Probably longer filter element life

7.5 Insulation and Heaters

For operation at temperatures above 250°F (121°C) the addition of hopper and wall heaters and insulation should be considered, especially if the process air or gas stream will go through the dew point frequently.

Insulation

If insulation is used, it probably will be advisable to use it on the entire dust collector surface. Where the process airstream will provide sufficient heat to maintain a temperature above the dew point, full surface insulation may be all that is needed. Usually a 3- to 4-in thickness of mineral wool or fiberglass is adequate for most locales and temperature differences. An external covering or jacket of thin aluminum or galvanized-steel sheet is recommended for protection of the insulation from rain and wind damage. Cost will range from around $5.50/ft^2$ for 3-in mineral wool and 18-gauge galvanized steel jacket installed; $5.75/ft^2$ for 3-in mineral wool and 0.040-in-thick aluminum jacket; to $6.75/ft^2$ for 4-in-thick fiberglass and 14-gauge HRS sheet. The R values will be about 6.7 for the 3-in mineral wool and about 20 for the 4-in-thick fiberglass.

Heaters

For additional temperature control, electric heater panels may be mounted on the external hopper or vertical wall surfaces of a dust collector prior to applying the insulation. Suppliers will provide calculated panel capacity in total kilowatts and provide a layout of panel locations as well as mounting studs and location templates. Supply voltage may be 115, 230, or 460 V ac, with larger jobs usually 460 V ac, three-phase, which should be loaded as evenly as possible on all three phases.

Thermostatic controls are supplied to maintain the desired temperature level. Junction boxes can be NEMA standard #4, #7, or #9 as required. Costs will run around $500/kW installed on the collector.

Heating capacity will depend on the user-specified ambient and airstream temperatures, how fast a start-up time to temperature is allowed, amount of insulation provided, and minimum temperature to be maintained in the collector.

7.6 Heat Exchanger

An air-to-air heat exchanger may be needed to cool the process airstream below the safe temperature of the filter elements. Hopefully the clean cooling air may be reused as a preheated combustion or makeup air supply to reduce costs.

A typical dusty airstream had a normal operating temperature of 300°F (149°C) inlet and 175°F outlet, with provision for upset conditions at 496°F inlet and 275°F outlet, and for extreme upset condition cooling from 800°F (427°C) inlet to 496°F (257°C) outlet. Cost will run around \$2.85 per ft^3/min not installed for a 37,800 ft^3/min (actual) airstream.

7.7 Lime or Filter Aid Injection

The injection of small volumes of pulverized limestone into the airstream is often used as a filter aid and/or sorbent for sulfur dioxide emission reduction. A filter aid powder injection system usually needs to deliver precisely metered small amounts over a delivery distance of 50 to 100 ft. The capacity will depend upon the airstream volume of the collector, but probably will run at 20 to 200 lb/h. Some small sizes may need to deliver capacities down to 2 lb/h (see Fig. 7.11).

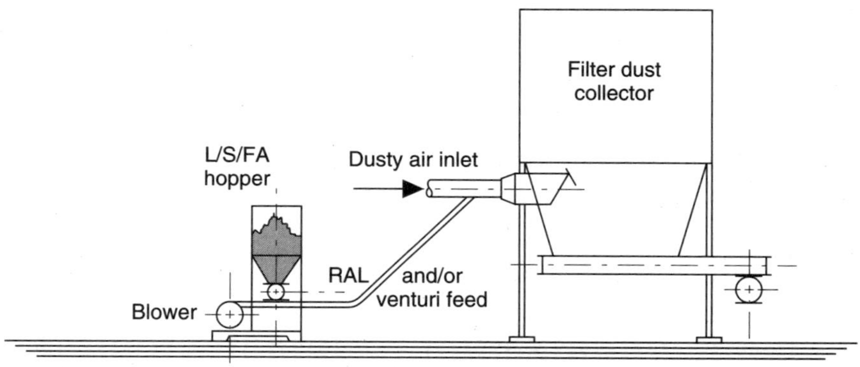

Figure 7.11 Lime, sorbent, filter aid injection: typical system.

A rotary air lock (RAL), venturi eductor, or vibratory feeder with air seal can be used to meter the small amount of powder volumetrically into a pneumatic conveyor which will join the dust collector inlet or duct ahead of the inlet to inject the powder into the airstream. A small hopper appropriate for the feed rate, a blower of 1 to 3 hp and 1½-in to 3-in-diameter line will also be needed. Total cost for such a system will run from about $4000 for the smaller to $6000 for the larger system, not installed. These injection systems are becoming more important as toxic pollutants must be dealt with.

7.8 Filter Precoat Powders

Initial coating of new filter bags with a lightweight powder that has a range of particle sizes will usually improve filtration performance. Tests have shown that many dust materials will begin to penetrate the filter medium immediately, especially with very fine particle sizes, unless a filter aid precoat powder is injected into the inlet.

By using pulverized limestone or a lighter-weight precoat, filters can be lightly coated prior to introduction of the dust itself. When a porous precoat layer of dust or filter aid is built up to condition the filter before normal dust loading operation is begun, tests have shown that both resistance to flow and the collection efficiency will benefit. Penetration of fine dust is reduced, efficiency starts at a higher level, and pressure drop or resistance across the filters is much lower. Filter bag life would also be lengthened. Lime has the additional benefit of being a sorbent for sulfur dioxide.

7.9 Hopper Flow Aid Devices

When difficult bulk solid materials will not flow out of hoppers, some help other than replacement with a better hopper can be used. There are numerous methods for aiding flow in dependable service, other than sledgehammers and poke rods, such as increasing hopper slope from 60° to 70°, use of a vibrating bin discharger described in Chap. 4, or use of an elongated or trough hopper with a screw conveyor discharge (see Fig. 7.12).

Fluidization

For powder or small-particle size dust, one of the best flow aids is some form of fluidization. Small 4- × 8-in air pads or fluidization panels may be easily added as needed to induce flow for many products. Compressed-air piping can be run outside the collector walls and the panels fed through a ½-in-diameter hole which also serves as

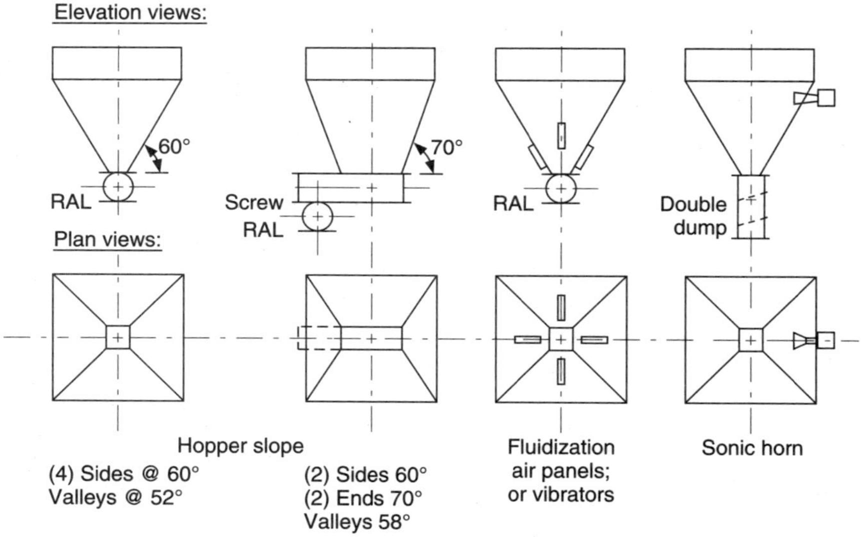

Figure 7.12 Hopper flow aid devices.

the support mounting with nuts on a threaded pipe connection to the air line. Airflow can be easily adjusted and timed or controlled to operate only when needed. Cost is low, around $25 to $50 per panel installed.

Vibrators

Electric or air vibrators are used very often to help induce flow from railcars and hoppers. But a word of caution is that a vibrator can pack the bulk solid tightly in many cases if overdone or if not regulated or controlled to provide just the action needed. Costs can run from a few hundred to several thousand dollars per vibrator.

Acoustic horns

These are a panacea for many operators these days, but again they may be overdone. Cost is in a range from a minimum of about $1200 to a maximum of about $6000 each. An additional benefit in many cases is the improvement in filter cleaning, and some collectors are being operated with just the acoustic horn for cleaning in place of a shaker or reverse-air cleaning system.

The beat goes on and developments occur, so watch for those that prove out and are justified by better performance and savings!

References

1. *Chemical Engineering,* Nov. 1992, p. 98.
2. *Chemical Engineering,* Oct. 1993, p. 17.
3. "5-MWe SNRB Demonstration Project," 1992, T. Gennrich and K. C. Hsieh, 3M Corp. and J. M. Wilkinson, P. Chu, and K. E. Redinger, Babcock & Wilcox.
4. U.S. EPA Specifications, #40 CFR, Part 60, Appendix B.
5. Anacon Inc., Anderson Inc.
6. Auburn International Inc., Babbitt Inc.
7. EPA, Research Triangle Park, Raleigh, NC.
8. Bindicator Inc., Binmaster Inc., Babbitt Inc., Monitor Mfg.
9. NFPA Procedure #68 and #69.
10. Gregory J. Black, *Chemical Engineering,* Feb. 1994, pp. 128–133.
11. BS & B, Fenwal, Fike, Oseco.
12. Mikropul Environmental Systems, William Gregg, "Improving Pulse Jet Dust Collector Operation: Effect of Gas Stream Inlet Design," *Powder and Bulk Solids Proceedings,* June 1990.
13. United McGill Corp., Fabric Filter Systems Catalog, May 1991.

Chapter

8

Dust Pickup and Conveying

The design purpose of dust pickup and conveying is to collect product or nuisance dust and to transport these dust products or bulk solids to an intended point of use or disposal as efficiently as possible. This is the same concept as an overgrown industrial vacuum cleaner.

8.1 Concept of Dust Pickup and Conveying Systems

Dust is picked up from a process point by suction into a hood or feed nozzle and is then conveyed through a duct to a collector. There the dust is separated from the airstream, the cleaned air is exhausted through the system fan, and the collected dust is delivered to its destination. Some systems use suction as above, and others use pressure to convey the dust. In another type of system a combination of negative and positive pressure is used on a material like sawdust which is sucked in, drawn through the fan, and then blown into a collector.

The dust capture hoods should collect as much of the process dust as possible to protect the breathing air for employees and the environment for everyone, as well as deliver the highest output yield of product. When transporting bulk solids or dust after capture, the input of solids to the duct is usually metered in by a volumetric feeder for uniform air-to-solids loading in the conveyor duct and for optimum conveying performance.

Dust capture velocity guidelines

The first design step is to select a capture velocity that will adequately collect the dust at its source; this can be determined by reference

TABLE 8.1 Range of Capture Velocities for Hood Designs

Method of dispersion of dust and contaminant	Examples	Capture velocity, ft/min
Released at practically no velocity into quiet air	Evaporation from tanks, degreasing, etc.	50–100
Released at low velocity into moderately still air	Spray booths, intermittent container filling, low-speed conveyor transfers, welding, plating, mixers, etc.	100–200
Active generation into zone of rapid air motion	Spray painting in shallow booths, rapid container filling, conveyor loading, crushers, screens	200–500
Released at high initial velocity into zone of very rapid air motion	Grinding, abrasive blasting, tumbling	500–2000

tables, past experience, or tests. OSHA guidelines and the *Industrial Ventilation* manual put out by the American Conference of Governmental Industrial Hygienists provide minimum capture velocities for many types of dust hoods and pickup inlets for typical equipment sources (see Table 8.1).

So capture velocities will range from 50 to 2000 ft/min depending on the dust particle size and density, motion of the air in the area, and hood or inlet design. With any particular dust a lower capture velocity can be used for nuisance dusts and low-air-motion sources. Higher pickup or capture velocities should be used for output sources that have a lot of air motion in the area and propel or throw the dust out, such as a grinding wheel, shaker screen, or drop chute into a container. Critical or hazardous dust products that must be most effectively and safely captured may require a higher capture velocity.

Hood and inlet design

The second essential design step is to develop an effective hood or inlet design and locate it as close as possible to the dust source for adequate collection. The capture velocity must be maintained across the full face area of the hood for effective pickup. A good average will not necessarily be effective enough, since each dust particle has to be captured by an adequate velocity.

When air is drawn into an opening, a flow pattern develops that typically converges downstream and just inside the opening. The minimum diameter of the airstream is smaller than the duct and is called the *vena contracta*. There is a loss of about 2 percent in static pressure that occurs as the airstream contracts, followed by a much greater loss

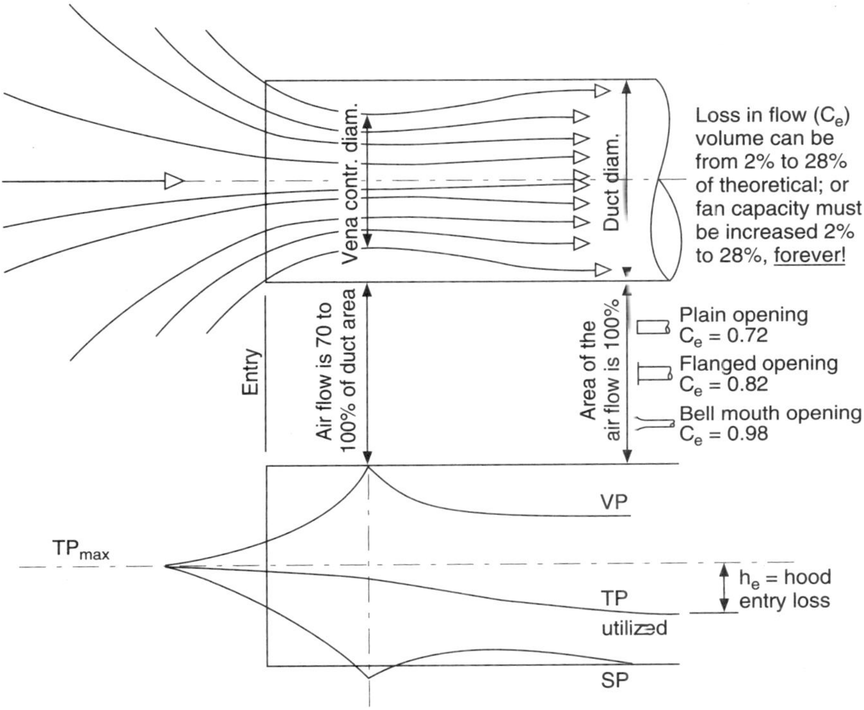

Figure 8.1 Vena contracta and airflow in hoods and inlets, with typical entry loss coefficients C_e.

as the airstream expands to the full diameter of the downstream duct. The area of the vena contracta varies with the shape of the inlet, but for most will range from 70 to 100 percent of the duct area. (See Fig. 8.1 and Ref. 1.)

The easiest way to reduce pressure drop is to use an optimum hood or inlet design that has a low entry loss H_e and a high coefficient of entry C_e. A plain duct opening has an entry loss of 0.93 of the velocity pressure (VP). A flanged opening has an entry loss of 0.49 of the VP (in WG). The flow or pickup efficiency which is known as the coefficient of entry C_e is the ratio of the actual rate of flow relative to the theoretical flow. The plain opening has a flow efficiency of 72 percent, and the flanged opening has an efficiency of 82 percent. Thus the flanged inlet opening has half the pressure loss and 10 percent higher flow volume.

While those are excellent gains, look at the bell mouth, or blunderbuss inlet. The flow efficiency rises to 98 percent, and the entry pressure loss decreases to 0.04 × VP. So just by going from a plain sharp-edged duct opening to a bell-mouth inlet, the efficiency of flow volume increases from 72 to 98 percent, and the pressure loss drops from 0.93 × VP all the way down to 0.04 × VP.

A square shoulder inlet to a booth or a dust collector enclosure, or an outlet, has an efficiency of flow of 82 percent and an entry pressure drop of 0.50 × VP, equivalent to the flanged opening. This is also known as a straight or direct inlet. By going to a tapered or flared inlet with an included angle of 30° to 60° in a round duct, the flow efficiency goes up to 95 percent and the entry loss drops to an average of 0.07 × VP. Finally by using a bell-mouth or rounded inlet, or out-

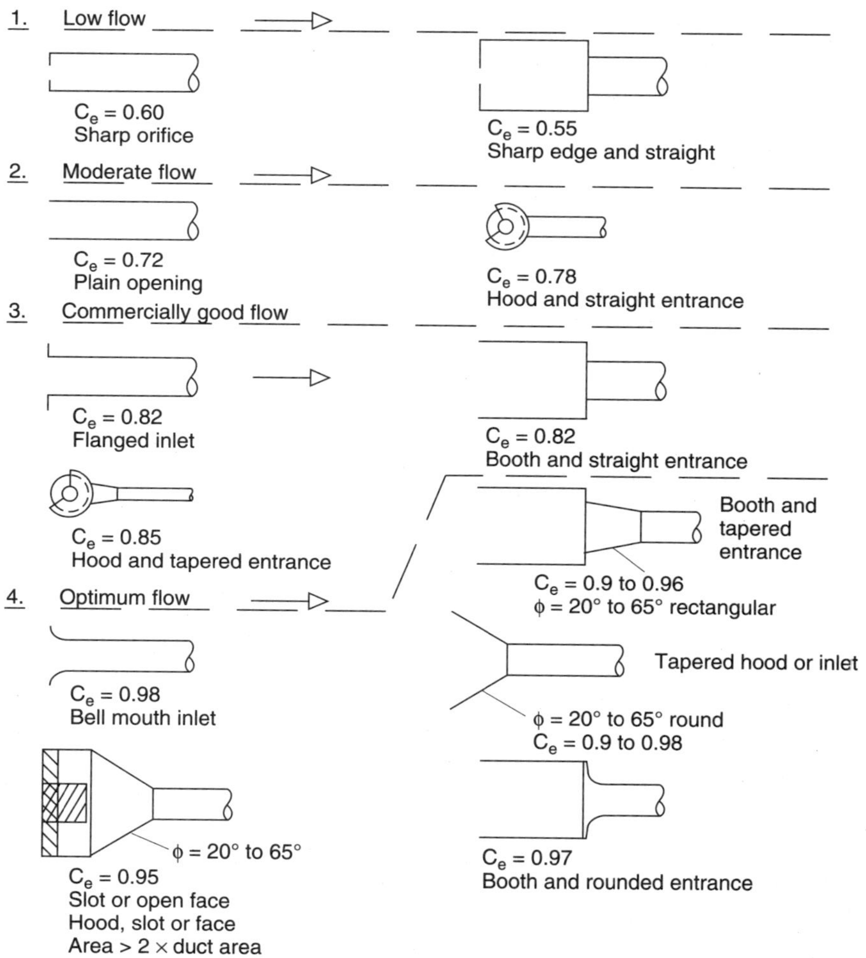

Figure 8.2 Hood and inlet design for efficient flow, with typical entry loss coefficients C_e.

let, the flow efficiency goes up to 97 percent and the entry loss drops to 0.06 VP. (See Fig. 8.2.)

A square shoulder inlet and outlet to a rectangular settling chamber or dropout box would have an entry loss of 1.5 × VP and a flow efficiency of 63 percent. This is quite poor but is exactly what the dropout chamber needs to get particles separated. So use these inlet and hood designs to do the job that is needed.

Cement dust pickup

A 250-ft/min velocity should be adequate to capture cement dust particles from kiln transfer and conveying points where high winds are not expected to disturb the collection. The total air volume required would be calculated based on the area of the pickup hoods, so if the design pickup area is 40 ft^2, then the air volume required is 250 ft/min × 40 ft^2 of openings = 10,000 ft^3/min. In this example, the hood design would allow a 6-in- or ½-ft-wide slot opening on all sides. This would permit 80 lineal ft of openings, since ½ ft × 80 ft = 40 ft^2 of open area.

$$\text{Dust pickup volume} = \text{capture velocity} \times \text{area of hood} = 250 \text{ ft/min} \times 40 \text{ ft}^2 = 10{,}000 \text{ ft}^3\text{/min} \tag{8.1}$$

$$\text{Pickup slot perimeter} = \frac{\text{hood area}}{\text{slot opening}} = \text{perimeter length (lineal ft)} \tag{8.2}$$

Thus the perimeter = 40 ft^2/0.5 ft = 80 lineal ft at a 6-in-high opening.

8.2 Dust System Sizing and Balancing

Proximity of inlet

The closer each pickup hood or inlet can be located to the dust source, the lower the total air volume required, since the same capture velocity is required but over a smaller opening area. Optimizing hood design for effective pickup at as close proximity to the dust sources as possible pays off in minimizing the air volume required. Working to achieve good inlet design without interference with the process results in the minimum size duct, collector, and fan, as well as a reduction in costs and emissions.

Energy costs

This minimizing is too often overlooked in inlet design, which means the entire system must then be overpowered to achieve adequate dust control. Thereby a higher initial investment along with a higher energy cost is incurred throughout the operating life of the system. Energy cost calculations will be covered in Chap. 9 [see in particular Eq. (9.1)].

Additional advantages of optimizing the dust pickup system accrue beyond the initial investment and operating costs for power:

- Fan size will be reduced, as will the drive motor.
- Reduced energy will lower power demand cost.
- The cumulative need for power plants will be reduced.
- The system will help reduce the world's total pollution level.
- Minimizing dust pickup air volume helps reduce plant makeup air required, and the associated heating or cooling energy needed.
- Reduced power needs will reduce plant wiring costs.

Some high-velocity low-volume exhaust systems use inlet velocities of 11,000 to 14,000 ft/min for short-distance localized dust pickup. These use a small conveying hose of up to 2½-in diameter or less, inlets located as close as possible to the source of dust generation, and centrifugal or positive displacement blowers capable of producing 12 to 20 in of mercury (160 to 272 in WG). The exhaust volume required can thus be minimized by using this higher pressure, with small-diameter inlets and short conveying distances. (See Table 8.2.)

TABLE 8.2 Sizing and Friction Loss in Round Ducts and Hoses

Diameter, in	Pressure drop, in WG per 100 ft, at velocity			Resistance* 90° elbow, equiv. ft of straight pipe	Airflow volume at 4500 ft/min, ft³/min
	3500 ft/min	4000 ft/min	4500 ft/min		
3	7.5	10.0	11.0	5.0	220
4	5.5	7.0	8.5	6.0	393
5	4.2	5.5	6.2	9.0	616
6	3.5	4.5	5.5	12.0	884
7	2.6	3.8	4.5	13.0	1,203
8	2.2	3.0	3.8	15.0	1,571
9	2.0	3.0	3.0	17.5	1,995
10	1.8	3.0	2.8	20.0	2,455
12	1.2	1.8	2.4	25.0	3,535
14	1.3	1.5	2.0	30.0	4,808
16	1.0	1.2	1.8	36.0	6,280
18	0.8	1.1	1.4	42.0	7,955
20	0.73	0.95	1.2	50.0	9,830
22	0.65	0.84	1.1	58.0	11,880
24	0.59	0.76	1.03	68.0	14,155
30	0.45	0.58	0.76	78.0	22,115

*For centerline radius of elbow = 1.5 × diameter of duct.

$$\text{Duct size} = \frac{\text{airstream volume}}{\text{conveying duct velocity}} \tag{8.3}$$

Thus,

$$\text{Size} = \frac{10{,}000\ \text{ft}^3/\text{min}}{4500\ \text{ft/min}}$$

$$= 2.22\ \text{ft}^2 \qquad \text{or 20-in-diameter duct}$$

The velocity needs to be high enough to safely convey the dust product, but not much higher so abrasion of the duct is minimized. In comparison, a 19-in diameter with an area of 1.97 ft^2 would require a velocity of 5075 ft/min, and a 21-in diameter with an area of 2.4 ft^2 would require a velocity of 4167 ft/min. Thus the 20 in with 2.1816 ft^2 would produce a velocity of 4584 ft/min for an optimum duct size.

Using Table 8.2, we can estimate the friction loss or pressure drop for a proposed dust collection duct system to an accuracy of around 10 percent using the *equivalent length* method. This is usually accurate enough for preliminary estimating for adding to the expected pressure drop in the collector to obtain a reasonable estimate of overall pressure drop for sizing the system fan. Where the duct system is more complex or when the fan size becomes critical later in the project, a more accurate detailed calculation can be made using the *velocity pressure* method, which can be found in numerous references such as the *Industrial Ventilation* manual.

Several reference sources, including *Industrial Ventilation,* provide ways to estimate duct system frictional pressure loss by either method. In many preliminary, and some final, project plans the less precise or ballpark system pressure drop may be calculated using the equivalent length method. As has been said about economic justification, "It is not always how detailed or minutely precise the calculation that is important; but the consistent use of a reliable and proven estimating basis will provide dependable guidance."[2]

For a 100-ft-long duct with four elbows and two pickup hoods, plus the filter collector, at a conveying velocity of 4500 ft/min, the friction loss can be estimated as follows from Table 8.2.

$$\text{Straight 20-in duct at 100 ft} \times \frac{1.2\ \text{in WG}}{100\ \text{ft}} = 1.2\ \text{in WG}$$

For elbow radius = 2.5 × diameter of duct:

$$\text{Elbows (4)} \times 0.34\ \text{in equivalent} = 1.36\ \text{in}$$

$$\text{Hoods (2)} \times 0.08 \times \text{VP of } 1.31 = 0.21 \text{ in}$$

$$\text{Collector estimated pressure drop} = 6.0 \text{ in}$$

$$\text{Total friction loss} = 8.77 \text{ in WG}$$

$$350°\text{F } (177°\text{C}) \text{ temperature correction } 1.53 \times 8.45 = 13.42 \text{ in WG.}$$

Thus the fan will need at least 13.5 in of static pressure suction to move the 10,000 ft^3/min of air at 4500 ft/min and 350°F.

Branch resistance. Using Table 8.2 ballpark estimates for each of the branch hoods and two 20-ft inlets can be calculated. Air volumes and pressure drops can be estimated, and duct areas and diameters calculated. If each branch collects 5000 ft^3/min, then an area of (5000 ft^3/min)/(4500 ft/min) = 1.1 ft^2, or 14-in diameter is needed for each branch duct. The main needs to be sized at

$$\frac{10{,}000 \text{ ft}^3/\text{min}}{4500 \text{ ft/min}} = 2.22 \text{ ft}^2 = 20\text{-in diameter}$$

If each branch is 20-ft long, then

$$(2) \times 2.0 \text{ in} \times \frac{20 \text{ ft}}{100 \text{ ft}} = 0.8 \text{ in WG} \qquad \text{(branch losses)}$$

$$80\text{-ft main} \times \frac{1.2 \text{ in}}{100 \text{ ft}} = 0.96 \text{ in} \qquad \text{main losses}$$

$$\text{Elbows} = 1.36 \text{ in}$$

$$\text{Hoods} = 0.21 \text{ in}$$

$$\text{Collector estimated pressure drop} = 6.0 \text{ in}$$

$$\text{Total friction losses} = 9.33 \text{ in WG}$$

$$\text{Temperature correction} \times 1.53 = 14.28 \text{ in WG}$$

Thus the branches add more pressure drop than the main, but in this case the total is very close to the same.

Duct system design

A conservative preliminary estimate based on realistic operating conditions is better than a detailed estimate made on optimistic or unknown operations. Obviously, the best final design is an accurate velocity pressure method based on final realistic operations. However, try a ballpark conservative method as outlined above for the preliminary estimate. When the project is firmed up, a different but known volume and layout is often the result. Then a final design and estimate can be made with the equivalent length or velocity pressure method. But always be sure that the basic data reflect known conditions and realistic expectations, and adjust the calculations when specifications change.

Duct layout, pickup points, volumes, and inlet designs. In Table 2.2 there are typical conveying velocities for the bulk materials listed. By comparison to similar materials, a velocity can be selected for unlisted materials. When firm specifications and bids are taken, then suppliers can run tests on the material and develop layout drawings for the system design. Inlet design types should be spelled out for the lowest entry pressure drops and highest flow coefficients and alternatives should be requested for optimum designs, performance, and cost.

Conveying duct size. Branch duct sizes can be determined for the necessary volumes and velocity using Eqs. (8.1) to (8.3), Table 8.2, and

1. Start with the branch duct farthest from the fan or dust collector. Size this branch for its required pickup volume and preferred conveying velocity. If the volume is 1200 ft^3/min and the velocity will be around 4500 ft/min, then from Table 8.2 select the diameter as 7 in.

2. Always maintain the same conveying velocity as closely as possible, in branches and mains with combined volumes! Thus if the second branch also requires a 1200-ft^3/min volume, it should also have a 7-in diameter. The main duct downstream from this combined volume of 2400 ft^3/min can have, from Table 8.2, a 10-in diameter and maintain the 4500-ft/min velocity.

3. If the third branch will bring in another 2400 ft^3/min, that branch should have a 10-in diameter. The subsequent main should handle 4800 ft^3/min and can have a 14-in diameter from Table 8.2.

4. If the fourth branch will add 5200 ft^3/min, it should also have a 14-in diameter. The final combined main duct must handle 10,000 ft^3/min and from Table 8.2, can have a 20-in diameter. The final velocity would actually be about (10,000 ft^3/min)/(2.1816 ft^2) = 4584 ft/min. This is close enough and will not add much friction.

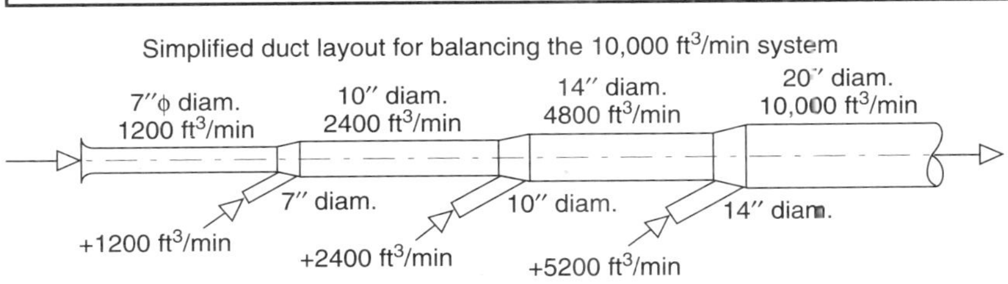

Figure 8.3 Balancing a duct system.

Fig. 8.3. As the branch lines join together, each succeeding duct run can be sized by the cumulative volume of the branches. The main line velocity can be designed to be slightly lower or higher than the branch velocities, up to 500 ft/min, which allows for balancing and adjusting the velocities and volumes. The final duct size carrying the full cumulative volume then runs into the fan or collector inlet. Remember that dust collector inlet and outlet performance can benefit by using high-efficiency transition designs, shown in Fig. 8.2.

8.3 Pneumatic Conveying

Most dust collectors for large-volume process and emission control are fed by dilute phase pneumatic conveyors. Dense phase conveyors are used more on delivery of bulk solids to and from transport vehicles, silos, blenders, and weigh hoppers. Where the dust product can be aerated easily, the fluidized conveyor or *Air-Slide* is frequently used. The operational parameters are compared briefly in Fig. 8.4.

A 7.3-ton/h capacity pneumatic conveyor comparison for the 10,000 ft^3/min portland cement collector discharge in Fig. 8.4 outlines several approaches. Duct and fan dilute phase, dense phase, and fluidization are the most common.

1. Duct and fan for 0.5 lb/in^2 gauge low-pressure ductwork

Conveying zone system	Diameter line	Velocity, ft/min	Air volume	Loading, lb/ft^3/min
A. Fluidized gravity [0.75 lb/in^2; 10 ft^3/(ft^2/min) = 10 ft/min air velocity]	6-in width	40	10 ft^3/ft	24.4
B. Duct and fan (14 in WG)	20-in diameter	4500	10,000 ft^3/min	0.0244
C. Dilute phase (6.4 lb/in^2)	3-in diameter	4480	220 ft^3/min	1.11
D. Screw feeder (10 to 30 lb/in^2)	3-in diameter	1650	80 ft^3/min	3.1
E. Dense phase tank (15 to 30 lb/in^2) and internal fluidization	3-in diameter	1240	61 ft^3/min	4.0

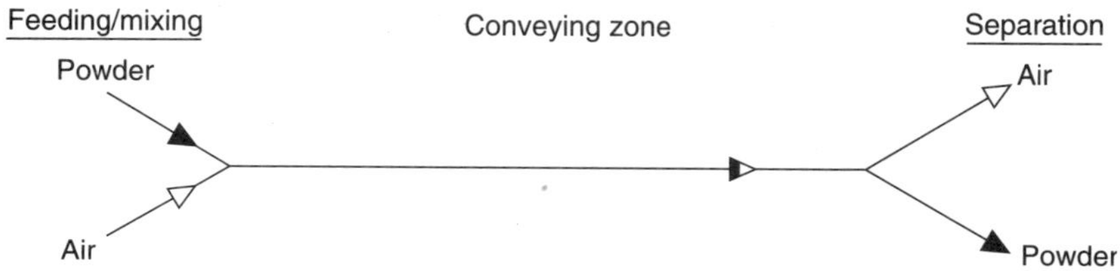

Figure 8.4 Pneumatic conveying system operational comparison (Capacity = 14,600 lb/h = 244 lb/min; cement density = 80 lb/ft^3).

$$\text{Conveying velocity} = 4500 \text{ ft/min}$$

$$\text{Loading} = \frac{41 \text{ ft}^3 \text{ air}}{\text{lb of dust}} = 0.0244 \text{ lb/ft}^3 \text{ air}$$

$$\text{Capacity} = 7.3 \text{ tons/h} \times \frac{2000 \text{ lb/ton}}{60 \text{ min/h}}$$

$$= 244 \text{ lb/min}$$

$$\text{Air volume required} = 244 \text{ lb/min} \times 41 \text{ ft}^3\text{/lb}$$

$$= 10{,}000 \text{ ft}^3\text{/min}$$

$$\text{Duct size} = \frac{10{,}000 \text{ ft}^3\text{/min}}{4500 \text{ ft/min}} = 2.22 \text{ ft}^2$$

$$\text{Use} = 20\text{-in diameter at } 2.1816 \text{ ft}^3$$

$$\text{Pressure drop} = 1.2 \text{ in WG per 100 ft} \qquad \text{(Table 8.2)}$$

2. Fluidized gravity for 0.75 lb/in^2 gauge and 2° to 5° slope

$$\text{Conveying velocity} = 100 \text{ to } 200 \text{ ft/min}$$

$$\text{Loading } 7.3 \text{ tons/h} = 244 \text{ lb/min dust}$$

$$\text{Volumetric loading} = \frac{244 \text{ lb/min}}{80 \text{ lb/ft}^3}$$

$$= 3.05 \text{ ft}^3\text{/min}$$

$$\text{Conveying chamber} = 6 \times 6 \text{ in} = 36 \text{ in}^2 \text{ area}$$

$$= 0.25 \text{ ft}^2 \text{ cross-sectional area}$$

$$\text{Velocity of dust} = \frac{3.05 \text{ ft}^3\text{/min}}{(0.25/3) \text{ ft}^2}$$

$$= 36.6 \text{ ft/min} \qquad \text{(for 2-in height of dust)}$$

$$\text{Air volume required} = 10 \text{ ft}^3\text{/ft}^2 \times \text{fluidized conveyor area}$$

$$= 10 \text{ ft}^3\text{/ft}^2 \times \frac{6\text{-in width} \times 12\text{-in length}}{144}$$

$$= 10\ \text{ft}^3/\text{ft}^2 \times 0.5\ \text{ft}^2/\text{ft length}$$

$$= 5\ \text{ft}^3/\text{ft conveyor length}$$

$$\text{Air flow velocity} = \frac{5\ \text{ft}^3/\text{ft}}{0.25\ \text{ft}^2\ \text{cross-sectional area}}$$

$$= 20\ \text{ft/min}$$

Conservative estimate doubles air volume required

$$\text{Air volume conservative} = 5\ \text{ft}^3/\text{ft} \times 2 = 10\ \text{ft}^3/\text{ft length}$$

$$\text{Air flow velocity conservative} = 20 \times 2 = 40\ \text{ft/min}$$

3. Dilute phase for 5 to 10 lb/in^2 gauge medium-pressure pneumatic conveyor

$$\text{Conveying velocity} = 4500\ \text{ft/min}$$

$$\text{Loading} = 0.9\ \text{ft}^3/\text{lb of dust} = 1.11\ \text{lb/ft}^3\ \text{of dust}$$

$$\text{Capacity} = 244\ \text{lb/min of dust}$$

$$\text{Air volume required} = \frac{244\ \text{lb/min}}{1.11\ \text{lb/ft}^3}$$

$$= 220\ \text{ft}^3/\text{min}$$

$$\text{Duct size} = \frac{220\ \text{ft}^3/\text{min}}{4500\ \text{ft/min}} = 0.049\ \text{ft}^2$$

(or 3-in diameter)

$$\text{Pressure drop} \approx 6.5\ \text{lb/in}^2\ \text{gauge}$$

4. Dense phase for 15 to 30 lb/in^2 gauge high-pressure pneumatic

$$\text{Conveyor} - \text{conveying velocity} = 1240\ \text{ft/min}$$

$$\text{Loading} = \frac{0.25\ \text{ft}^3}{\text{lb of dust}} = \frac{4.0\ \text{lb}}{\text{ft}^3\ \text{of dust}}$$

$$\text{Capacity} = 244\ \text{lb/min of dust}$$

$$\text{Air volume required} = \frac{244 \text{ lb/ft}^3}{4.0 \text{ lb/ft}^3} = 61 \text{ ft}^3\text{/min}$$

$$\text{Duct size} = \frac{61 \text{ ft}^3\text{/min}}{1240 \text{ ft/min}} = 0.049 \text{ ft}^2$$

(or 3-in diameter)

$$\text{Pressure drop} \approx 22 \text{ lb/in}^2 \text{ gauge}$$

Tests should be made prior to final sizing of all conveyors for each application.

References

1. *Industrial Ventilation,* 19th ed., 1986, pp. 4-4 to 4-12.
2. T. C. Davis and H. L. Hessler, "How the Du Pont Company Appraises Its Performance," presented to the American Management Association, Du Pont Co., 1949, p. 8.
3. *Industrial Ventilation,* pp. 6-39 to 6-41.

Chapter

9

Fans and Airflow

The air or gas in any dust collection system must be propelled by a fan or other means of air movement. By far the most commonly used prime mover in industrial filter dust collection systems is a centrifugal fan, at operating static pressures up to 25 in WG and for airstream volumes up to 100,000 ft^3/min. Higher delivery volumes and pressure ratings are available. Operating temperatures up to 500°F (260°C) service have been normal in dust collection, but higher temperatures require fan construction that is available for up to 1000°F (538°C). Higher temperature ranges from 1000 to 1200°F are under trial for industrial furnace and boiler (IFB) applications, kiln exhaust, electric power plants, and in combination SO_x, NO_x, Rox, Box (SNRB) absorbent and filtration applications.

For higher pressures of 20 to 40 in WG, a narrow bladed centrifugal fan can be used efficiently for volumes to 20,000 ft^3/min. Pressures to 70 in WG and volumes to 50,000 ft^3/min are available. Where a pressure blower with 1.0 to 3.0 lb/in^2 gauge, or 27.7 to 83.1 in WG, is needed at low volumes for aeration, fluidizing, or pneumatic conveying, either the radial blade or ring compressor type of centrifugal blower is suitable.

For pneumatic conveying pressure requirements of 3.0 to 15.0 lb/in^2 gauge, a multistage radial centrifugal pressure blower is capable of pressures up to 10 lb/in^2 gauge, but a positive displacement blower is usually selected for the 5.0- to 15.0-lb/in^2 pressure ranges. For air compressor service the positive displacement reciprocating or screw type is used for pressures from 15.0 to 150 lb/in^2 gauge and above.

Vane axial or propeller-type fans are rarely used for dust collection since they have a low static pressure rating in the 1- to 5-in WG range. The vane axials are often used where a high-volume capacity at low pressure is needed in localized dust or spray hood exhaust, and the propeller type is often used for general area ventilation and exhaust. In-line centrifugal fans can save space in ductwork, in dryers or ovens, or other special applications for air movement or for pressure boosting.

9.1 Centrifugal Fan Efficiency

The simplest centrifugal design is the straight radial blade, or paddle wheel, fan, and it generally has the lowest initial cost. The radial blade fan is the workhorse for exhausting dirty air through the fan since the blades are self-cleaning to a large extent. This is the fan design chosen for handling stringy, abrasive, agglomerating, or fibrous material, or where there is a heavy dust load or large particle sizes drawn through the fan. Efficiency is relatively low, in the 60 to 75 percent range.

Higher efficiency can be achieved with a backward inclined flat blade centrifugal fan. Clean air or a light load of fine dust may be handled with efficiencies of 75 to 80 percent. Initial cost is a little higher than with the radial blade, but energy cost savings usually make them worthwhile. This fan is most often used on the clean-air side of a dust collector.

The centrifugal fan with an airfoil shape and a backward curved blade has the highest efficiency, is quieter, and does not pulsate in operation. However, it is the most expensive and is subject to abrasive wear on the hollow blades. This airfoil design is usually only justified where the highest efficiency is needed for energy savings, or for extremely high airstream volumes up to 300,000 ft^3/min.

Since a large proportion of dust collection applications are on the clean side anyway, the backward inclined flat blade fan is most widely used for its reduced energy and horsepower needs. Further advantages are lower speed and noise, a nonoverloading horsepower curve, and a stable pressure curve. Their increased cost over radial blade fans can usually be justified by the operating cost savings. (See Fig. 9.1.)

9.2 Fan Drives

There are three main types of fan drives:

- V-belt–drive fans are most commonly used because of the flexibility of speed changes. An increase in fan speed can be made quickly to increase the airstream volume or the static pressure capacity of the fan, or both, when within the fan and motor rating.

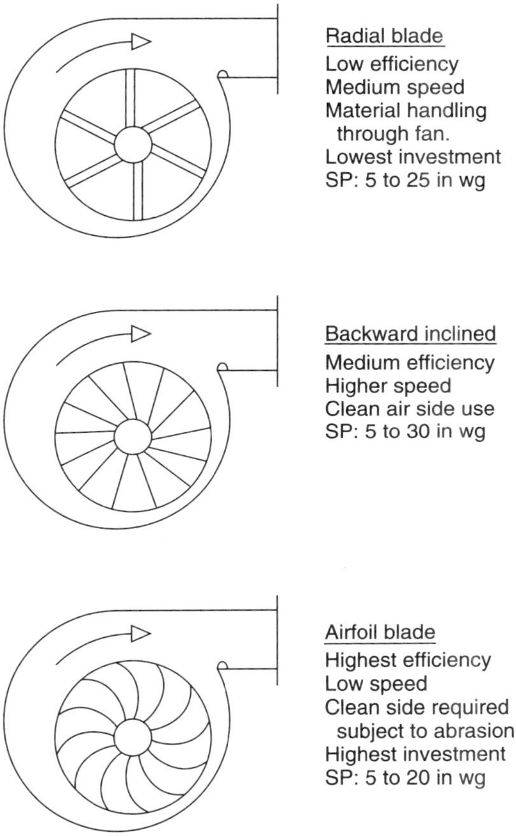

Figure 9.1 Fan blade designs and efficiency, with typical operating factors.

- Direct-drive fans should be considered as an alternative, however, since there are no belt transmission losses. This increases efficiency by around 7 percent, thus reducing energy and horsepower needs, and delivers a more constant output speed and volume. So initial investment cost savings can accrue from reduced horsepower and elimination of the belt drive, guard, and bearings, plus the power cost savings because of the higher efficiency. Where the ratings fit and the speed is not likely to need changing, direct-drive fans produce savings and should be considered.
- Variable-speed drives are finding more application on fans for the ability to quickly adjust air volume delivered in proportion to the process load, combustion rate, or absorbent injection rate needed.

Speed controllers can be programmed to limit starting torque, peak starting current, time to reach full speed, or respond to operating loads by feedback signals. As dependability increases and costs come down, these variable-speed controls should be considered more often for large fans, since they have already proven themselves on fractional to 5-hp motor drives on feeders and other small applications.

9.3 Fan Selection Considerations

Some important items to consider when selecting a fan are

1. Volume of the total airstream in cubic feet per minute determined from the hood and ductwork pickup volumes and resistance per Chap. 8, or from process volume requirements.
2. Fan static pressure required for the ductwork system.
3. Clean or dirty side fan location at the collector.
4. Type of dust or bulk solid material to be handled, and whether it is corrosive, flammable, or explosive in nature.
5. Temperature, altitude, and humidity of operation.
6. Noise limits required.
7. Optimum fan rating selection for size and horsepower, direct or V-belt drive suitability for maximum efficiency, and whether variable-speed control has advantages.
8. Corrosive materials' effect on the choice of fan materials of construction, from carbon steel, to stainless steel or other alloys, spark-resistant construction, fiberglass, polyvinyl chloride (PVC), or coated.
9. Spark-resistant fan construction as classified by the Air Movement and Control Association (AMCA) may be needed to handle explosive or flammable materials in three types as follows:

   Type A. All parts in contact with the air or gas must be nonferrous material.

   Type B. The fan shall have an entirely nonferrous wheel and ring about the opening through which the shaft passes.

   Type C. The fan construction shall not permit two ferrous parts of the fan to rub or strike.

9.4 Air Density Correction Factors

Correction factors for altitude, temperature, and humidity can be calculated using the *Industrial Ventilation* manual procedures, Section

6-22, under "Corrections for Temperature, Moisture, and Elevations." As they point out, "correction for temperatures between 40 and 100 F and/or elevations between − 1,000 and + 1,000 feet are seldom required with the permissible variations in usual exhaust system design."[1]

The *Industrial Ventilation* manual further notes a major reference point in fan laws:

> It is helpful to remember that a centrifugal exhauster connected to a given system will exhaust the same VOLUME regardless of air density. The mass, or weight, of air moved, however, will be a function of the DENSITY, as will the pressure developed and power consumed.

Humidity or moisture correction is well covered in Section 6 of the *Industrial Ventilation* manual and is less frequently encountered in

TABLE 9.1 Fan Altitude and Temperature Correction Factors

Air temperature, °F	Altitude, ft										
	0	500	1000	1500	2000	2500	3000	3500	4000	4500	5000
0	0.87	0.89	0.91	0.92	0.94	0.96	0.98	0.99	1.01	1.03	1.05
40	0.94	0.96	0.98	1.00	1.02	1.04	1.06	1.08	1.10	1.12	1.14
70	1.00	1.02	1.04	1.06	1.08	1.10	1.12	1.14	1.16	1.18	1.20
80	1.02	1.04	1.06	1.08	1.10	1.12	1.14	1.16	1.19	1.21	1.23
100	1.06	1.08	1.10	1.12	1.14	1.16	1.19	1.21	1.23	1.25	1.28
120	1.09	1.12	1.14	1.16	1.18	1.20	1.23	1.25	1.28	1.30	1.32
140	1.13	1.15	1.18	1.20	1.22	1.25	1.27	1.29	1.32	1.34	1.37
160	1.17	1.19	1.22	1.24	1.26	1.29	1.31	1.34	1.36	1.39	1.42
180	1.21	1.23	1.26	1.28	1.30	1.33	1.36	1.38	1.41	1.43	1.46
200	1.25	1.27	1.29	1.32	1.34	1.37	1.40	1.42	1.45	1.48	1.51
250	1.34	1.36	1.39	1.42	1.45	1.47	1.50	1.53	1.56	1.59	1.62
300	1.43	1.46	1.49	1.52	1.55	1.58	1.61	1.64	1.67	1.70	1.74
350	1.53	1.56	1.59	1.62	1.65	1.68	1.72	1.75	1.78	1.81	1.85
400	1.62	1.65	1.69	1.72	1.75	1.79	1.82	1.85	1.89	1.93	1.96
450	1.72	1.75	1.79	1.82	1.86	1.89	1.93	1.96	2.00	2.04	2.08
500	1.81	1.85	1.88	1.92	1.96	1.99	2.03	2.07	2.11	2.15	2.19
550	1.91	1.94	1.98	2.02	2.06	2.10	2.14	2.18	2.22	2.26	2.30
600	2.00	2.04	2.08	2.12	2.16	2.20	2.24	2.29	2.33	2.38	2.42
650	2.10	2.14	2.18	2.22	2.26	2.31	2.35	2.40	2.44	2.49	2.54
700	2.19	2.23	2.27	2.32	2.36	2.41	2.46	2.50	2.55	2.60	2.65
750	2.28	2.33	2.37	2.42	2.47	2.51	2.56	2.61	2.66	2.71	2.76
800	2.38	2.43	2.48	2.52	2.57	2.62	2.66	2.72	2.76	2.81	2.86

Note: Fans are rated at standard air conditions of 0.075 lb/ft^3 density at 70°F temperature and sea level; therefore, operating pressures must be corrected to standard conditions before fans are selected from manufacturers' rating tables or curves.[2]

Caution: Motors must be sized for the highest density condition (lowest fan factor) or lowest temperature and altitude at which it is expected to start up or operate. Pressure at operating conditions × fan correction factor = pressure at standard [Eq. (9.1)]. Operating horsepower × fan correction factor = standard horsepower [Eq. (9.2)]. Also, motor size must allow for belt drive losses of around 7 percent from available horsepower at the fan, if a V-belt drive is used.

dust collection systems, so it will be left for reference to that manual. However, fan correction for altitude and temperature effects on air density must be made frequently, so it will be explained herein using fan manufacturers' tables (see Table 9.1).

9.5 Fan Correction Factors

Fan correction factors for altitude and temperature are listed in most fan manufacturers' catalogs, similar to Table 9.1. Referring to the footnote in Table 9.1, pressure at operating conditions × fan correction factor = pressure at standard of 70°F and sea level. Thus for a fan with 10 in WG operating pressure at 300°F (149°C) temperature at 2000-ft altitude, with a fan correction factor from Table 9.1 of 1.55:

$$\text{Operating pressure} \times \text{fan corrector factor} = \text{standard pressure}$$

$$10\ \text{in} \times 1.55 = 15.5\ \text{in WG} \qquad (9.1)$$

The fan horsepower standard can be found similarly:

$$\text{Operating horsepower} \times \text{fan correction factor} = \text{standard horsepower}$$

$$20\ \text{hp (for example)} \times 1.55 = 31.0\ \text{hp} \qquad (9.2)$$

Case Study Thus for a dust pickup and collection system operating at 10 in WG at 300°F (149°C) at an altitude of 2000 ft with a fan horsepower of 20 hp, the above correction of fan requirements must be made before an appropriate fan size can be selected from the fan manufacturer's rating tables. This correction accounts for an ambient temperature start-up and the location's altitude.

Obviously the installation's altitude remains the same but must be accounted for to arrive at the standard pressure rating. And the fan will have to be sized for 70°F (149°C) ambient start-up unless the process allows the temperature to come up prior to starting air movement, or some load limiting start-up controller is used.

So, instead of choosing a fan for 10 in WG and 300°F (149°C) operation at 2000 ft with a 20-hp drive, the fan must be sized for 15.5 in WG at 70°F (21°C) start-up. The fan manufacturer's rating table would call for 31.0 hp at these conditions, which will require the use of a 40-hp motor to adequately pull this load at start-up.

This fan selection and its cost could be explored with suppliers for a more efficient fan type, such as an airfoil, possibly with a direct drive to eliminate the ±7 percent belt drive losses, and/or a soft-start controller to limit the maximum current required at start-up. The increased initial cost may be offset by the savings in installation and annual operating costs for a fan requiring say 27 to 29 hp at start-up, which with a direct drive would keep the required motor under 30 hp. The more efficient fan would still deliver the required airstream volume at 10 in WG at 300°F (149°C) operating conditions and at 15.5 in WG for ambient start-up.

9.6 Density Correction

Density correction for temperature, altitude, and humidity above standard conditions will cause the air density to decrease, as noted above. The density factor d from the *Industrial Ventilation* manual, figure 6-20, Air Density Correction Factor, is the ratio of actual to standard density at the design operating conditions. The air density correction factor is 0.65 for 300°F (149°C) and 2000-ft altitude.[3]

Note that d is the inverse of the fan correction factor in Table 9.1.

$$\begin{aligned} d &= \frac{1}{\text{fan correction factor}} \\ &= \frac{1}{1.55} \qquad \text{for 300°F, 2000 ft} \\ &= 0.65 \end{aligned} \tag{9.3}$$

The standard density of 0.075 lb/ft³ for air at 70°F and sea level should be divided by the fan correction factor from Table 9.1 for altitude and temperature, to get operating density:

$$\begin{aligned} \text{Operating density} &= \frac{\text{standard density}}{\text{fan correction factor}} \\ &= \frac{0.075 \text{ lb/ft}^3}{1.55} \\ &= 0.048 \text{ lb/ft}^3 \end{aligned} \tag{9.4a}$$

Using the density correction factor directly provides the same operating density.

$$\begin{aligned} \text{Operating density} &= \text{standard density} \times d \text{ air correction} \\ &= 0.075 \text{ lb/ft}^3 \times 0.65 \\ &= 0.048 \text{ lb/ft}^3 \end{aligned} \tag{9.4b}$$

Reduced duct friction loss at lower densities

At elevated temperatures or higher altitudes, the duct friction loss is reduced because of the lower air density. The duct friction charts used previously for equivalent length static pressure loss per 100 ft are based on standard air density. For other densities due to temperature, altitude, or humidity, the friction loss for standard air should be multiplied by the density correction factor. This determines the true friction loss for air at other densities. So for a 4-in WG friction loss, for example, calculated at standard conditions, the density correction or the fan correction factor can be used to get operating loss at 300°F (149°C) and 2000-ft altitude:

$$\begin{aligned}\text{Operating duct friction loss} &= \text{calculated friction loss at standard density} \\ &\quad \times d \\ &= 4 \text{ in} \times 0.65 \\ &= 2.6 \text{ in WG} \qquad (9.5a)\end{aligned}$$

$$\begin{aligned}\text{Operating duct friction loss} &= \frac{\text{calculated friction loss at standard density}}{\text{fan correction factor}} \\ &= \frac{4 \text{ in}}{1.55} \\ &= 2.6 \text{ in} \qquad (9.5b)\end{aligned}$$

9.7 Limiting Drive Power

Drives must start up a fan moving ambient or cold temperature air in most cases, and this requires more power than when operating with higher-temperature lower-density air. Even when operating at or near ambient conditions a damper or slow-start type variable-speed, torque control may be advantageous for both start-up and process control. This can be handled in one of several ways:

1. Size the motor for the higher horsepower at cold start-up.
2. Install a damper at the inlet or outlet of the fan to limit the airstream volume either manually or automatically at start-up; then open it up gradually to full volume when at operating temperature.
3. Install a variable-speed, variable-torque, current, or power load control to ramp up gradually to full speed and torque at operating temperature, and as an overload control.
4. The damper or drive controller can be adjusted manually or automatically in response to process feedback signals for optimizing, or control of, capacity, air volume, pressure, combustion efficiency, or other needs.

The second and third methods allow the use of a smaller motor and drive, and their cost should be evaluated for justification. Larger motor horsepowers can be handled with dampers at least cost, while smaller motors up to around 5 hp may be handled with the variable-speed or torque controls economically. Cost savings on the motor, electric wiring and controls, plus demand charges for higher start-up amperage, as well as process control, can make fan dampers or motor controls very advantageous.

9.8 Fan Laws

A fan operating at a fixed speed can have an infinite number of ratings along its characteristic curve. The operation can vary from *free delivery,* or essentially no static pressure which a propeller exhaust or ceiling fan could have, to its highest static pressure rating where output volume would be at a minimum.

When a fan must operate by pulling air through a duct system and a dust collector and blow air out an exhaust stack or back into a plant, then the fan rating will become fixed at the system resistance or a set pressure level on the fan curve. This resistance can change if the airstream volume is changed by damper opening, collector pressure differential, or speed control. But in normal operation a process or dust collection system fan will operate at a fixed pressure and volume level on the fan curve at a fixed speed.

As stated in the *Industrial Ventilation* manual,

> One characteristic of the fan is of prime importance—when operated at a fixed RPM in a steady state system (no changes to hoods, ductwork, fittings, or to damper or collector back pressure), the fan is a constant volume machine. For example, even though temperature and consequently density of air entering a hood is changed, the fan will continue to move or capture the same volume of air.

Once a fan is built the volume developed by the fan can only be changed by changing the speed or the resistance, which is the operating pressure.[1]

The ductwork, dampers, type collector, and exhaust configuration will determine the system resistance and thus affect volume levels. But within a fixed system setting, the fan speed, or resistance and pressure, are the only controls on fan delivery. At a new speed, the operating point will be determined by the basic fan laws given in Table 9.2.

These basic fan laws apply only to the fan. For application to a system, they would apply only where pressure losses of all components vary as the square of the velocity through them. So the fan laws would apply to hoods, inlet and outlet transitions, duct resistance, acceleration and deceleration losses, and to centrifugal types

TABLE 9.2 Basic Fan Laws

Variable	Speed change	Density change
Volume	Varies directly with speed ratio $CFM_2/CFM_1 = RPM_2/RPM_1$	Volume does not change
Pressure	Varies with square of speed ratio $P_2/P_1 = (RPM_2/RPM_1)^2$	Varies directly with density $P_2/P_1 = D_2/D_1$
Horsepower	Varies with cube of speed ratio $HP_2/HP_1 = (RPM_2/RPM_1)^3$	Varies directly with density $HP_2/HP_1 = (D_2/D_1)$

of dust collectors. However, pressure losses through filter collectors and some wet-type collectors increase *directly* with the change in air volume.

$$\text{New collector pressure drop} = \text{original pressure drop} \times (\text{percentage volume change} + 100\%) \tag{9.6}$$

Thus when airstream volume increases 20 percent through a filter collector with 5-in WG pressure drop average, the new pressure drop will be 5 in WG × 1.2 = 6 in WG.

For a ducting system, the pressure drop increases as the *square* of the speed ratio, or the equivalent in the square of the volume change.

$$\text{New duct pressure drop} = \text{original pressure drop} \times (\text{percentage change} + 100\%)^2 \tag{9.7}$$

Thus when the volume changes by 20 percent, the pressure drop in the ductwork and hoods increases by the square of the change, or for a 5 in WG drop in the ducting, the new pressure drop will be 5 in WG × (1.2 × 1.2) = 5 × 1.44 = 7.2 in WG.

$$\text{Fan brake horsepower} = \frac{\text{volume (ft}^3\text{/min)} \times \text{total pressure (in WG)}}{6356 \times \text{mechanical efficiency of fan}} \tag{9.8}$$

For the 20 percent change in speed, the volume increases by 10,000 × 1.2 = 12,000 ft^3/min, and the pressure drop increases from 5 to 7.2 in WG, so the horsepower increases due to both changes:

$$\text{BHP 1} = \frac{10{,}000\ \text{ft}^3\text{/min} \times 5\ \text{in}}{6356 \times 0.7} = 11.24\ \text{brake hp (bhp)}$$

$$\text{BHP 2} = \frac{12{,}000\ \text{ft}^3\text{/min} \times 7.2\ \text{in}}{6356 \times 0.7} = 19.42\ \text{bhp}$$

Horsepower for a fan or ducting system can be calculated directly since it varies as the cube of the fan speed or volume increase. Thus, from Table 9.2,

$$\text{New horsepower} = \text{original horsepower} \times (\text{original speed} + \text{change})^3 \quad (9.9)$$

For the 20 percent change then,

$$\text{New horsepower} = 11.24 \text{ bhp} \times (1.2)^3 = 11.24 \times 1.728 = 19.42 \text{ bhp}$$

which is the same as Eq. (9.8) calculation above.

9.9 Fan Installation Guidelines

Fan ratings are based on ideal conditions, and field installation may not measure up to the ideal unless carefully planned. Some of the factors to look out for are listed here.

Duct system

As previously discussed, the duct system can make a big difference in overall pressure drop, so watch for leakage and the most efficient inlets, hoods, and duct sizing, as well as the optimum conveying velocity without overkill. Also joint connection and pipe surface smoothness can affect pressure drop or friction, especially on long runs. Roughness correction factors should be covered with the suppliers and a standard of performance settled by reference to *Industrial Ventilation,* figure 6-19.[3]

Fan inlet

The inlet should maintain a straight flow into the fan (see Figs. 9.2 and 9.3). Limited installation space is the chief culprit in forcing short radius elbows to be connected directly to the fan inlet. A 2 to 4 times duct diameter distance of straight duct just prior to the inlet will reduce losses 20 to 30 percent, and 8 to 10 times the diameter distance will reduce inlet losses 45 to 80 percent.

If there is not sufficient space, then turning or straightening vanes, an inlet damper with a similar effect as the straightening vanes, and an inlet box with vanes or egg crate straighteners should be considered, and all would reduce inlet losses considerably. A square duct 90° elbow directly connected to the fan inlet without vanes will have a volume loss (ft^3/min) of 17 percent, one short 90° vane drops that to 8 percent, and three 90° vanes in the elbow will reduce the loss to 4 percent.

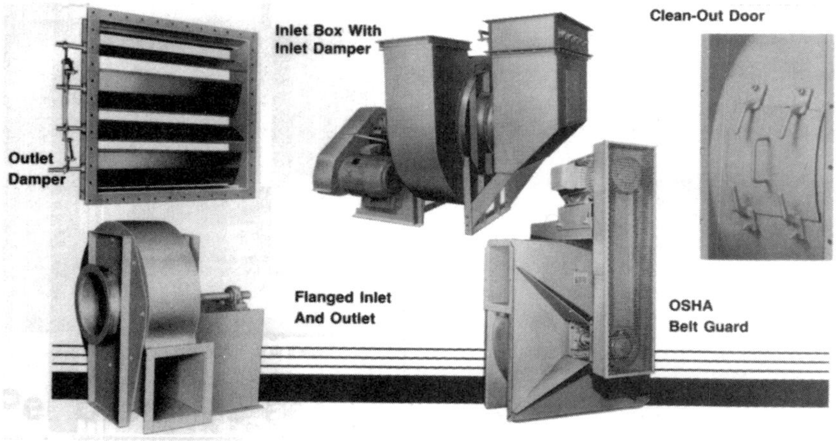

Figure 9.2 Fan inlet and outlet designs.

The fan static pressure increase required for a 90° elbow without vanes at the fan inlet is 45 percent, with one vane the static pressure increase drops to 18 percent, and with four 90° vanes the static pressure increase is reduced to 9 percent.

Fan outlet connection designs

These designs usually increase the system resistance. Elbows on the discharge increase the static pressure loss due to the normal

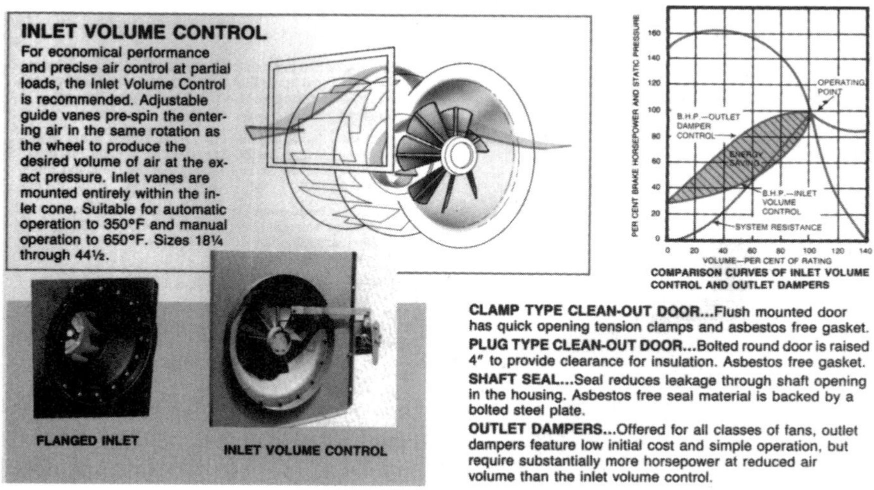

Figure 9.3 Inlet volume control.

nonuniform fan discharge velocity and the inherent resistance to turning. A short-radius elbow or a sharp 90° elbow on the discharge creates a loss of 1 × velocity pressure (1 VP), where VP is determined by the conveying velocities at standard density and where velocity is given by

$$V = 4005\sqrt{\text{VP}}$$

Rearranging,

$$\text{VP} = \left(\frac{\text{velocity}}{4005}\right)^2 \tag{9.10}$$

Typical losses would be 1 in WG at 4005 ft/min, 1.26 in WG at 4500 ft/min, 1.56 in WG at 5000 ft/min, 1.89 in WG at 5500 ft/min, and 2.24 in WG at 6000 ft/min.[1]

When the fan outlet discharges directly to the atmosphere, with no exhaust duct or stack, it produces only half as much loss, or 0.5 VP, as a short elbow. A straight exhaust stack of the same size as the fan outlet and two diameters long or more reduces the loss to zero.

The Evasé exhaust stack is a flared or gradual enlargement of the stack at a 10° included angle, or 5° angle on each side. The purpose is to reduce the air discharge velocity efficiently so the velocity pressure can be regained. The practical discharge velocity from an Evasé outlet is about 2000 ft/min. Lengthening the Evasé outlet beyond the point of reaching 2000 ft/min gives diminishing returns. As an example, for a volume of 8300 ft^3/min a 20-in-diameter exhaust has a velocity of 3800 ft/min. When flared at a 5° angle on each side, the velocity will reach 1940 ft/min in a 28-in outlet in a length of 46 in. So a length of two to three times the diameter is the maximum needed.

If a higher velocity is desired for vertical dispersion of heated or contaminated air, then the Evasé outlet is probably not justified, since a velocity of 3000 to 3500 ft/min or more is recommended. However, the use of contaminated air dispersion as a means of pollution control is on the way out with the tighter emission limits and available technology to filter or neutralize most contaminants rather than dispersing them to the atmosphere.

Fan location considerations

Some benefits can be gained just by careful planning of an installation, such as:

1. Locate the fan after the dust collector wherever possible, so only clean air is drawn into the fan. This minimizes material buildup,

TABLE 9.3 Fan Rating and Selection Table

[22¼] WHEEL DIAMETER: 24 in.
OUTLET AREA: 2.90 sq. ft.
MAXIMUM BHP = 1.94 (RPM\1000)^3 TIP SPEED, fpm = 6.28 × RPM

	OV	½" SP		1" SP		1-½" SP		2" SP		2-½" SP	
CFM	FPM	RPM	BHP	RPM	BHP	RPM	BHP	RPM	BHP	RPM	BHP
2900	1000	563	.34	681	.59						
3190	1100	594	.41	705	.67	807	.96				
3480	1200	627	.48	732	.76	828	1.06				
3770	1300	662	.56	761	.85	852	1.17	938	1.51		
4060	1400	699	.65	791	.96	878	1.30	960	1.66	1038	2.03
4350	1500	737	.76	822	1.08	906	1.43	984	1.81	1059	2.20
4640	1600	776	.88	855	1.21	935	1.58	1009	1.97	1081	2.38
4930	1700	815	1.01	888	1.36	965	1.73	1037	2.15	1105	2.57
5220	1800	855	1.16	923	1.52	996	1.91	1065	2.33	1131	2.78
5510	1900	896	1.32	959	1.70	1027	2.09	1095	2.54	1159	3.00
5800	2000	937	1.50	996	1.89	1060	2.31	1125	2.76	1187	3.23
6380	2200	1019	1.91	1073	2.34	1129	2.78	1188	3.25	1247	3.75
6960	2400	1102	2.38	1152	2.86	1201	3.33	1254	3.83	1309	4.35
7540	2600	1185	2.94	1232	3.46	1277	3.96	1325	4.49	1374	5.04
8120	2800	1269	3.57	1313	4.14	1355	4.69	1398	5.24	1443	5.81
8700	3000	1353	4.30	1395	4.92	1435	5.51	1474	6.09	1515	6.69
9280	3200	1438	5.13	1477	5.79	1515	6.43	1552	7.05	1589	7.68
9860	3400	1523	6.06	1560	6.77	1596	7.45	1631	8.12	1666	8.78
10440	3600	1608	7.11	1644	7.86	1678	8.59	1711	9.30	1744	10.00
11020	3800	1693	8.27	1727	9.06	1760	9.84	1792	10.59	1823	11.34

	OV	6" SP		7" SP		8" SP		9" SP		10" SP	
CFM	FPM	RPM	BHP	RPM	BHP	RPM	BHP	RPM	BHP	RPM	BHP
6380	2200	1614	7.66								
6960	2400	1657	8.49	1748	9.76						
7540	2600	1704	9.40	1791	10.74	1876	12.11	1958	13.50		
8120	2800	1756	10.38	1839	11.80	1920	13.24	1999	14.71	2076	16.21
8700	3000	1811	11.45	1890	12.94	1967	14.45	2043	16.00	2117	17.57
9280	3200	1869	12.62	1945	14.17	2019	15.76	2091	17.38	2162	19.02
9860	3400	1929	13.90	2002	15.51	2073	17.16	2143	18.85	2211	20.56
10440	3600	1991	15.30	2062	16.96	2130	18.68	2197	20.43	2263	22.21
11020	3800	2054	16.82	2123	18.54	2190	20.31	2254	22.12	2318	23.97
11600	4000	2120	18.48	2186	20.25	2251	22.08	2313	23.94	2375	25.86
12180	4200	2188	20.29	2250	22.10	2313	23.98	2374	25.91	2434	27.87
12760	4400	2257	22.24	2316	24.10	2376	26.03	2436	28.01	2494	30.04
13340	4600	2329	24.36	2385	26.27	2442	28.24	2499	30.27	2556	32.35
13920	4800	2403	26.63	2455	28.59	2509	30.61	2564	32.69	2618	34.82

SOURCE: Chicago Blower Corp., Glendale Heights, Ill., "Centrifugal SQB Fans," Bulletin SQB-106, July 1987.

CLASS I	RPM 2112
CLASS IS	RPM 2374
CLASS II	RPM 2668

3" SP		3-½" SP		4" SP		4-½" SP		5" SP	
RPM	BHP	RPM	BHP	RPM	BHP	RPM	BHP	RPM	BHP
1130	2.60								
1151	2.80								
1172	3.01	1237	3.46						
1196	3.23	1258	3.70	1319	4.18				
1220	3.47	1281	3.96	1340	4.45	1397	4.96		
1247	3.72	1305	4.23	1362	4.74	1418	5.26	1472	5.80
1303	4.28	1358	4.81	1411	5.36	1463	5.92	1515	6.49
1363	4.90	1415	5.47	1465	6.05	1514	6.65	1562	7.25
1425	5.61	1475	6.21	1522	6.82	1569	7.45	1615	8.09
1490	6.41	1537	7.03	1583	7.68	1627	8.33	1671	9.00
1557	7.31	1601	7.96	1645	8.63	1688	9.31	1730	10.01
1628	8.33	1668	8.99	1709	9.69	1750	10.40	1791	11.13
1701	9.45	1738	10.15	1776	10.86	1815	11.60	1853	12.35
1777	10.70	1811	11.42	1846	12.16	1882	12.92	1918	13.69
1854	12.07	1885	12.82	1918	13.58	1951	14.37	1935	15.17

11" SP		12" SP		13" SP		14" SP		15" SP	
RPM	BHP	RPM	BHP	RPM	BHP	RPM	BHP	RPM	BHP
2190	19.16	2260	20.77						
2233	20.69	2301	22.38	2368	24.09				
2278	22.31	2345	24.07	2410	25.86	2474	27.66	2536	29.48
2327	24.03	2391	25.86	2454	27.72	2516	29.60	2577	31.50
2380	25.85	2441	27.76	2502	29.70	2562	31.65	2621	33.63
2435	27.80	2494	29.78	2552	31.78	2610	33.81	2668	35.86
2492	29.88	2549	31.92	2606	33.99	2662	36.09		
2551	32.10	2607	34.20	2661	36.34				
2611	34.47	2666	36.63						

abrasion, erosion, and the resulting unbalancing of the fan, and provides for utilization of more efficient fan designs.

2. Eliminate all elbows and other inlet obstructions as much as possible (see above discussion on inlets, and ductwork).
3. Where practical, arrange the discharge and rotation of the fan so that the discharge is in the final direction desired. This eliminates the need for further elbows after the fan.
4. Locate the fan where an Evasé outlet can be used for minimum pressure drop.
5. Locate the fan for easy service and inspection access.
6. Easy access permits checking for incorrect rotation after service or if vibration problems appear.

9.10 Energy and Power Cost Calculations

Saving energy never goes out of style! Ask the Japanese about current benefits. After you reduce the pressure drop through the ductwork, minimize filter collector static pressure drop through good design, and minimize fan horsepower; then the calculation of energy costs is fun, because you can appreciate the money saved in initial investment and in continuing operating and energy costs.

For the 10,000 ft³/min volume and 12- to 14-in WG pressure drop, the horsepower can be calculated from Eq. (9.8):

$$\text{Fan brake horsepower} = \frac{\text{volume} \times \text{total pressure}}{6356 \times \text{mechanical efficiency}}$$

$$= \frac{10{,}000 \text{ ft}^3/\text{min} \times 13 \text{ in WG}}{6356 \times 0.7}$$

$$= 29.22 \text{ bhp}$$

For a practical installation the motor should be sized for cold start-up, with alternative consideration of a direct drive if possible, damper or variable-speed torque load controller if justified, and with adequate horsepower for the 300°F (149°C) operation at 2000-ft altitude. But for simplicity, let's adjust horsepower for a 1.55 fan correction factor for temperature and altitude: Fan brake horsepower = 29.22 × 1.55 = 45.29 bhp at start-up, per Eq. (9.2).

Energy cost will be based on the power cost per kilowatt, say $0.05/kW-h:

TABLE 9.4 Estimated Fan Costs (1994)

Volume, ft³/min	Static pressure, in WG	Temperature, °F	RPM	Motor, hp	Price, $	Ratio $/(ft³/min)
262	8	70	3,450	0.75	500	1.91 (Dir)
400	8	70	3,310	1.5	850	2.13
550	11	250	3,675	5.0	1,600	2.91
550	11	70	3,185	3.0	1,900	3.45 (B, Ex)
800	5	70	2,246	2.0	1,100	1.38
1,000	6	70	2,515	3.0	1,500	1.50
1,000	12	125	3,613	5.0	1,650	1.65 (Ch)
1,100	12	120	3,576	7.5	1,700	1.55
1,500	6	70	2,060	5.0	1,700	1.13
1,500	20	70	5,056	7.5	2,100	1.40
2,000	10	120	2,773	7.5	1,600	0.80
2,000	12	250	3,285	10	1,700	0.85
2,500	8	70	1,675	7.5	2,200	0.88
3,000	10	70	1,946	10	2,000	0.67
4,000	10	200	2,303	15	2,700	0.68
4,000	11	200	1,854	15	3,000	0.75 (VS)
5,000	10	95	1,814	15	2,650	0.53
5,000	12	70	2,575	15	2,650	0.53
5,000	13	70	1,654	20	3,000	0.60
6,000	20	70	2,194	40	5,500	0.92
6,500	10	70	2,758	20	2,650	0.41
8,000	9	70	3,143	20	3,600	0.45
8,500	17	150	2,092	50	5,100	0.60
10,000	10	70	2,546	25	4,200	0.42
10,000	15	150	1,605	40	5,100	0.51
12,000	12	110	1,082	40	6,100	0.51
15,000	12	70	2,190	40	5,000	0.33
19,400	9	70	1,519	40	5,600	0.29
20,000	20	70	2,488	100	9,900	0.50
21,500	10	70	1,775	60	6,500	0.30 (Dir)
24,000	14	70	2,229	75	7,600	0.32
24,000	14	70	1,750	75	6,820	0.284 (Dir)
24,700	21	70	1,083	150	13,300	0.54
26,500	17	70	2,154	125	9,600	0.36
28,000	10	230	1,978	100	9,000	0.32
32,000	10	300	1,789	100	10,500	0.33
40,000	17	70	1,600	150	19,300	0.48
45,000	11	70	1,266	125	14,300	0.32
50,000	12	150	1,467	150	14,900	0.30
50,000	12	375	1,722	200	16,000	0.32
63,000	12	150	1,343	200	16,300	0.26
80,000	18.8	70	1,192	300	33,000	0.42 (I, Ev)

Notes: B = spark resistant class "B" AMCA Spk. Res Cl. B; Ch = chemical and mill duty motor; Ex = Expl. Prf.; VS = variable-speed drive; Dir = direct drive; I = inlet damper; Ev = Evasé exhaust stack. Included in all above fans: flanged inlet and outlet drain plug, access door, weather cover guard, manual outlet damper, and totally enclosed fan-cooled (TEFC) motor.

$$\text{Power cost} = \text{brake horsepower} \times 0.746\ \text{kW/hp} \times \$0.05/\text{kW-h}$$
$$\times \left(50\ \frac{\text{weeks}}{\text{year}} \times 7\ \frac{\text{days}}{\text{week}} \times 24\ \frac{\text{h}}{\text{day}}\right)$$
$$= 45.29 \times 0.746 \times \$0.05/\text{h} \times 8400\ \text{h/year}$$
$$= \$14{,}190\ \text{per year} \tag{9.11}$$

Suppose the required brake horsepower could be reduced about 10 percent. Then the total brake horsepower would be 45.29 − 4.53 = 40.76 hp. From Eq. (9.11), power cost = 40.76 × 0.746 × ($0.05/h) × 8400 h/year = $12,771 per year, or a savings of $1,419 per year. That can quickly justify some additional investment in a higher efficiency fan or reducing the horsepower required for operating or start-up. (See Table 9.3.)

9.11 Estimated Fan Costs

Based on fan purchases in recent years from major reputable manufacturers, the estimated 1994 costs are presented in Table 9.4.

9.12 Importance of the Fan

Next to the collector itself, the fan is usually the most important and costly piece of equipment in a collection system, unless the ductwork is extensive. It will incur the greatest annual operating costs in many cases, primarily in energy required. Thus there is much benefit in the careful specification and selection of the fan in overall collection system performance and cost.

References

1. *Industrial Ventilation,* pp. 6-22, 6-42.
2. Chicago Blower Corp., "Engineering Guide," Glendale Heights, Ill. 60137, p. EG-1A.
3. *Industrial Ventilation,* pp. 6-44, 10-14.
4. Chicago Blower Corp., "Centrifugal SQB Fans," Bulletin SQB-106, Glendale Heights, Ill., July 1987.

Chapter

10

Electric Controls

Electric controls are used for normal start-stop functions and increasingly to improve the filtration effectiveness and efficiency on dust collectors. They are able to automate the cleaning system operating sequence, transmit operating parameter signals to centralize reporting, and integrate process equipment control. A familiarity with some of these options will permit their selection for the best overall operation and efficiency.

The National Electrical Code (NEC) is developed and revised every 3 years by the National Fire Protection Association (NFPA). The 917-page 1993 edition defines and summarizes the current minimum requirements for safeguarding persons and property from the hazards of using electricity. As a help in explaining and understanding the NEC guidelines and safe practice for wiring and code compliance, the NFPA publishes an NEC handbook.

10.1 Electric Control and Wiring Systems

While not everyone concerned with dust collection needs to be an electrical expert, it is valuable to know where to refer to the mother lode of wiring information in the United States. The national codes in other countries and local codes as well as company practice will also influence appropriate wiring and safety practice.

In the United States, enclosures are rated by Underwriters' Laboratories (UL), NEC, and NEMA standards. The Canadian Standards Association (CSA) has ratings similar to those of the United States. DIN is the Deutschland Ingineering Normalization. The International Electrotechnical Commission (IEC) sets ratings on test methods similar to UL and CSA but with slightly different performance requirements. VDE is the Verband Deutscher Electrotechniker.

One single international standard would seem to be much more productive. However, we need to be aware of these various standards, since they are found frequently in equipment that we buy and use now.

The primary use for electric controls on filter collectors is to provide automatic on-demand cleaning of filters at preset pressure drop levels across the filters. Electric controls predominate in use rather than pneumatic. The benefits in cleaning and operation are

1. Maintaining lower pressure drop across the filters
2. Permitting the use of reduced air pressure in cleaning
3. Improving the average filtration effectiveness level by controlling filter cake thickness as desired
4. Reducing penetration of particles into the filter
5. Lengthening filter life
6. Lowering the consumption of compressed air for cleaning
7. Lowering overall energy consumption by reducing the fan and cleaning air power required

Another use of electric controls is to interconnect and sequence the operation of accessory equipment with the collector for best operation. Some typical aims at start-up would be

1. Start the discharge conveyor or valve prior to starting the collection system fan.
2. Start the filter cake precoat process, prior to or in conjunction with collector cleaning, and fan start-up.
3. Then start the fan and normal input feed of dust, in some cases after the process air flow has reached a temperature greater than the dew point.

Some aims at shutdown would be

1. Stop the input dust feed and system fan first.
2. Stop the filter cleaning after 5-10 min.
3. Stop the discharge equipment last, after another 5 to 10 min, to clear dust from the hopper.

Of course, other equipment in the process can be interlocked for the best or intended control at any point in the operation, such as

1. Diverter valve or damper actuators can be opened or closed for cleaning or flow volume control.

2. The collector can be bypassed until temperature is above dew point.
3. For maintenance service on one module.

This sequencing of operations can automatically achieve what good manual control would do and can usually do it better and faster. On start-up there would be no dust feed until the collector cleaning and discharge equipment are in operation, or after a desired dust or precoat cake is built up on the filters. At shutdown the filter cleaning and the hopper discharge equipment may need to be run after the fan and dust feed are stopped, to ensure that no large quantity of dust will remain in the collector after normal operation. Of course, a power outage or equipment failure may interrupt the normal sequence, but that calls for the operator's action anyway.

A photohelic control will provide the signal to initiate on-demand cleaning as recommended above. The photohelic control will indicate the differential pressure drop across the filters like a magnahelic gauge but will also provide pressure sensing with adjustable set points for a high- and a low-pressure level. The photohelic control comes with a contact output at each set point level for automatic cleaning control.

Another route can use a magnahelic gauge for pressure indication and a pressure sensing transmitter for sending a 4- to 20-mA, or other, signal to a remote location. The photohelic is also available with a transmitter output for a 4- to 20-mA, or other, output.

A solid-state timer is used to provide electric pulse outputs at adustably timed intervals to the solenoid and diaphragm valves that deliver the pulse jets of air or gas to clean the filters. So that the best cycle time for cleaning doesn't have to be guessed at, a photohelic control can be added to the timer to provide the basic on-demand cleaning operation that is standard equipment on most continuously operating filter collectors. The timer is typically set to provide a 0.005- to 0.01-s-long electric pulse for each cycle. The interval between cycles during on-demand cleaning can be from a few minutes to an hour or more, depending on the dust characteristics and process conditions.

Contact or transmitter signal outputs from the differential pressure control add another option for remote or integrated control systems. Selector or push-button switches and indicator lights permit easy change from manual, to timed, to automatic on-demand cleaning or for manual operation of accessory equipment. All these controls can be incorporated in a single enclosure such as NEMA #4, 7, 9, or 12, as needed to suit the exposure conditions.[1]

Manual switches may be included in the panel enclosure for other functions, such as duct or fan damper actuators. These can be wired to respond automatically to a volume or temperature process signal, or to the photohelic control's differential pressure level. For instance, the fan air volume may be modulated by inlet or outlet damper control to provide the optimum burner or process operating level, or opened or shut for off-line cleaning of one module at a time. Obviously such functions may also need to be controlled with manual switches for start-up or emergency conditions.

Level controls may be wired into the panel to signal or actuate other functions such as alarms, or interlocked to collector equipment. Emissions monitors may signal to the panel for alarms or action. As long as the cost does not become excessive for custom wiring, it should be beneficial to include all control functions needed in the original equipment. Then compatibility would be assured and optimum start-up and operation facilitated. Future troubleshooting can also be more effective with full instrumentation.

Variable-speed controls can be interlocked to permit adjustment of operating capacity of inlet dust or sorbent feeds. A long-standing use is for adjustment of fan output volume and pressure in response to process needs. Many variable-speed controls can also ramp or adjust the torque delivered, and the power or input amperes drawn. Some have several preset speed levels that can be selected for known outputs.

10.2 Programmable Logic Controller (PLC)

One of the best and most significant control developments of the last several decades is the PLC. Actually a dedicated computer with input and output terminals for industrial control, the PLC is very well adapted to plant installation and programming. It also replaces a hard-wired panel of timers, relays, and other devices.

A big advantage of the PLC is that changes in the circuit, including the addition of more functions in the operation, are easily made in the program. Instead of going out to a supplier for another timer or relay, a programmer can add these additional functions in just a few minutes.

Another advantage is their ready interconnection with personal computers for control and data recording. Many plants have maintenance technicians trained in the programming and installation of PLCs, to support the engineers and management who have adopted these tools. The PLCs' speed of response in milliseconds and adaptability are hard to beat. Just in cost and space savings, the use of PLCs in place of hard-wired panels will quickly pay for themselves.

"PLCs often are the solution to the excessive maintenance and downtime caused by relay based controls. They can boost machine performance by adding functions that are difficult to provide with relays."[2] They are easy to troubleshoot, usually cost less in the long run, and are smaller in size. In addition, the existing program is easy to document. If you have ever tried to track down a "lost" wiring diagram that was supposed to be kept inside the electric control panel door, you can appreciate the documentation advantage.

Cost savings and more precise control have been the driving forces in the adoption of this new technology. Engineering universities and technical colleges are now offering electronic control and PLC courses that overlap mechanical and chemical courses.

Standardized programming languages have been developed by the IEC in conjunction with NEMA, so check them out before jumping in if you have not already adopted one. The IEC standard #65A includes four programming languages, and a new standard has been issued recently:

1. Ladder logic diagram
2. Sequential function chart
3. Instruction list
4. Structured text
5. IEC standard 1131-3 for programmable controller languages, issued by IEC, and under study in the United States as a NEMA standard

IEC standard 1131-3 was developed so that users could extend programs to include new data types, functions, and function blocks. This language will also provide a common file format so that programs can be exchanged in the future between different systems. This would permit software to be independent of the PLC hardware and thus be easily transferred. There is also another standard, IEC SC 65C/WG6, being developed to permit this language's function blocks to be applied in distributed control systems. The future use of PLCs should grow and broaden as these and other developments are available.[3]

As a Weyerhauser Company systems engineer testified,

> It's hard to beat the speed and repeatability of programmable controllers for operating gates, valves, motors, etc. As programmable controllers become more intelligent, the two technologies (PCs and PLCs) are merging. Today, our PLCs talk with our PCs, Personal Computers. Sophisticated PLCs can handle serial inputs and perform other PC like

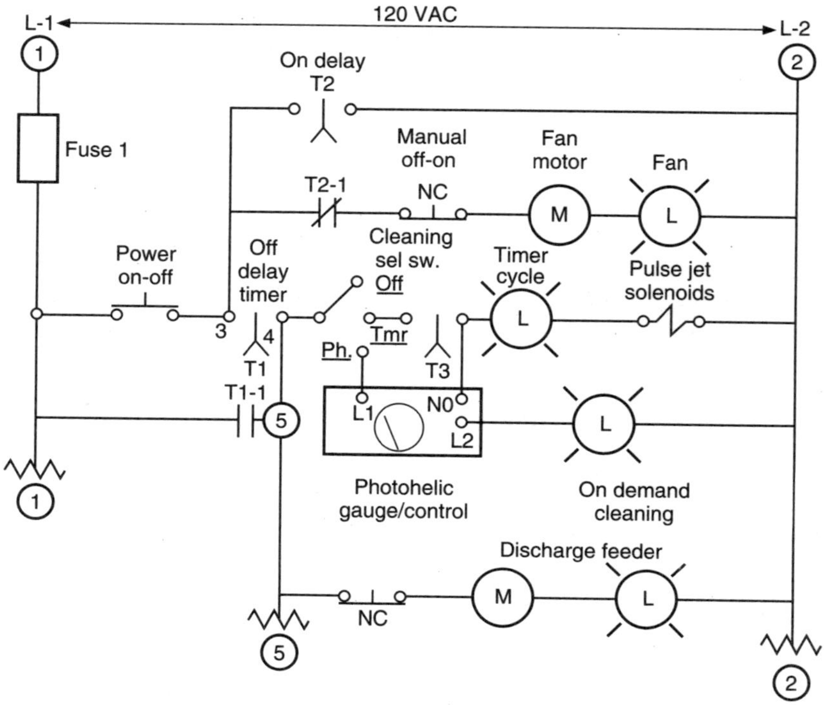

Figure 10.1 Pulse-jet collector control wiring, typical diagram. Options: "on-demand" auto cleaning; timer cycle cleaning; "off" = no cleaning; fan delays starting after cleaning starts; fan manual shut off; off delay timer keeps discharge and cleaning on; and manual discharge switch off.

functions. Tomorrow's PLCs will have many of the capabilities and functions of a PC.[4]

A special report on PLCs in *Instrumentation and Control Systems* magazine[5] says that, "Users want systems integration services, distributed control system/PLC and PC/PLC marriages, alternative languages, and the list goes on."

Programmable logic controllers are now able to perform proportional + integral + derivative (PID) control loop functions, using PID standards of the Instrument Society of America (ISA). This capability and other recent developments provide

- Better process control of temperature, pressure, flow, and level.
- Motion control of acceleration and deceleration, and the ability to sense and transmit position movement accurately.

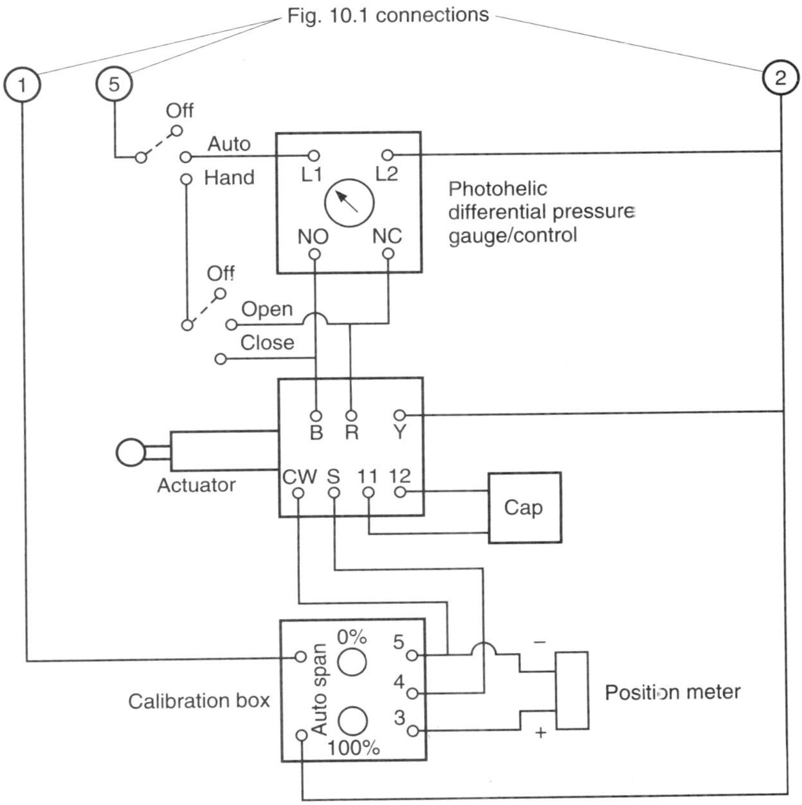

Figure 10.2 Fan damper actuator control and photohelic—added.

- Distributed control systems (DCS) which make network communication possible between a number of PLCs. All are supervised by a larger PLC or host computer, link onto a network like Ethernet, and run other discrete input-output (I/O) functions along with PID loops.

For all that, when the chips are down, the typical technician and engineer wants a small, simple PLC, with ladder logic programming, that is expandable, in a dependable enclosure, at a lower overall cost than hard-wired panels, and without the complexity of computer systems. Just remember, the goal is cost-effective dust collection that beats the pollution regulations and emission limits. So future improvements must be accomplished through present or easily available and understandable technology.

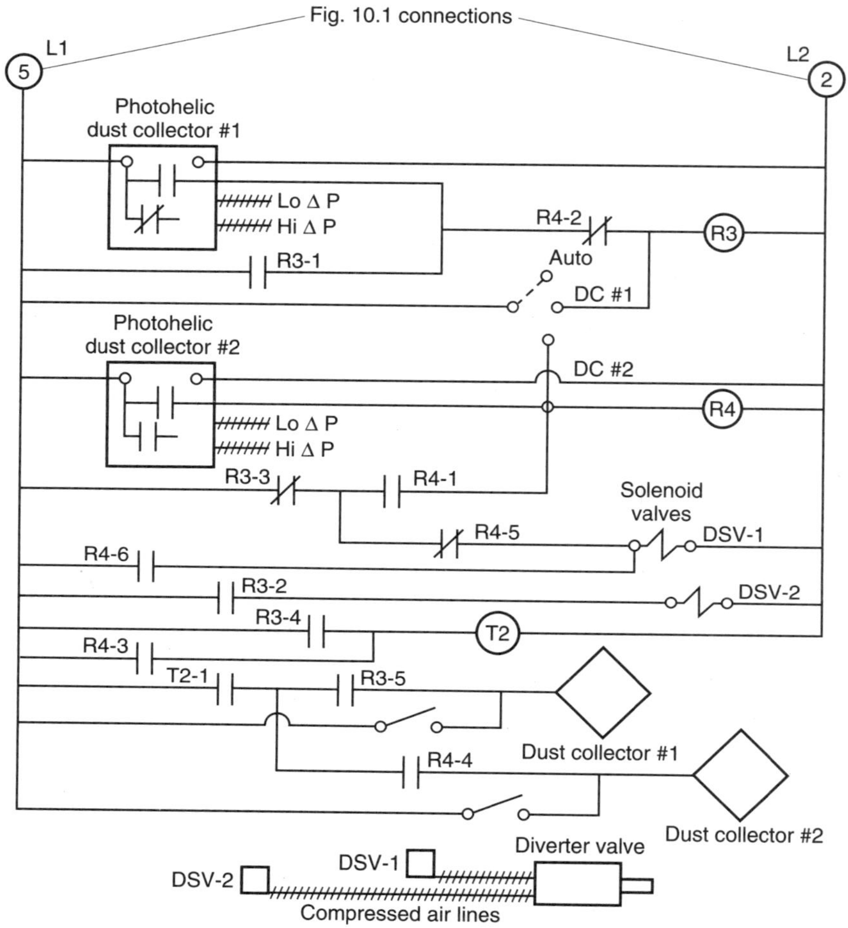

Figure 10.3 Diverter valve control added—on-demand cleaning, alternating collectors off-line.

10.3 Typical Control Wiring Diagrams and Costs

Costs for control panels can vary widely. Two big factors are the number of functions included and the type of enclosure. But for preliminary estimating, some ballpark costs are

1. Simple two–push-button, one three-way selector switch for timer or photohelic control for on-demand cleaning, and NEMA standard #4 enclosure (Fig. 10.1): $800.00.

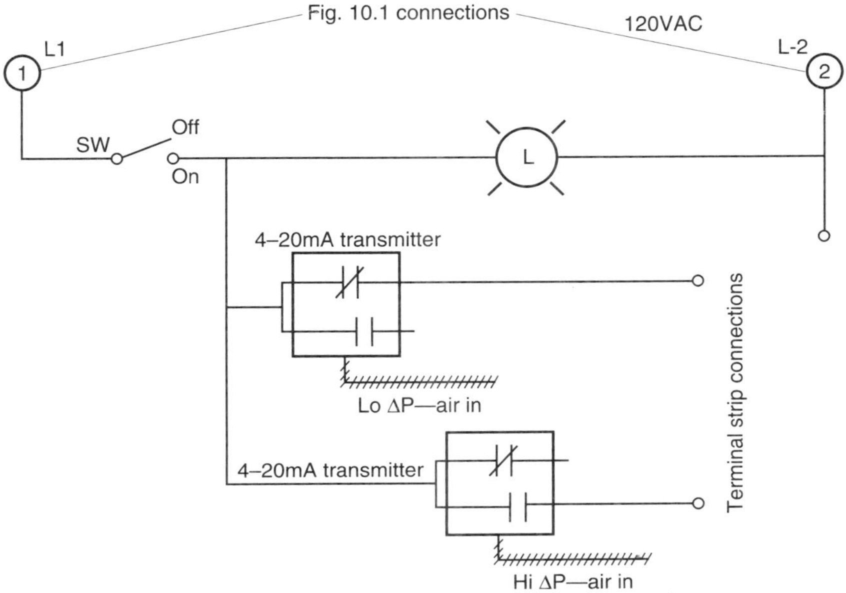

Figure 10.4 Multiple-pressure transmitters, for remote indication and control, added.

2. Addition to Fig. 10.1 of another photohelic control and fan damper actuator control, calibration box, and position meter (these costs are separate) (Fig. 10.2): $1,100.00.
3. Diverter valve control added for alternating collectors off-line and for on-demand cleaning when on-line (diverter valve costs separate) (Fig. 10.3): $3,500.00.
4. Similar to number 3, but with addition of two differential pressure transmitters for remote indication of high and low pressure (Fig. 10.4): $4,500.00.

References

1. National Electrical Manufacturers Association.
2. *Plant Engineering,* Aug. 27, 1987.
3. *Instrumentation and Control Systems,* Oct. 1993, p. 73.
4. *Plant Engineering,* March 12, 1987.
5. *Instrumentation and Control Systems,* March 1993.

Chapter

11

Installation, Operation, Testing, and Troubleshooting

For all industrial filter dust collector applications, an overall plan should cover a successful installation, operation start-up, performance tests, and provide a basis for troubleshooting. This overall plan should be included in the specification to better ensure the expected results in operation and performance. This chapter will explain how it ought to work. A typical multiple collector installation in progress is shown in Fig. 11.1.

All the preparation and design work culminates in achieving successful operation of a filter dust collector. The proof of the pudding comes in meeting or exceeding the design operating goals. The last part of a successful project is planning the installation from the ground up, literally having; the best start-up procedures; and using the most reliable performance testing for optimizing results.

11.1 Installation

Foundation

If the location will be outdoors, the soil strength needs to be checked to be sure it is adequate for the foundation loading. Usually a 3000-lb/ft^2 strength is adequate for all but unusual weight, wind, or seismic loading. For an existing building slab or structural system, weights and loadings must be checked for adequate available foundation strength.

Foundations will need to be designed so that they impose a lower loading than the soil strength. Something around 1500-lb/ft^2 loading on a 3000-lb/ft^2 soil should be safe. This allows the use of 3000 lb/in^2 concrete for piers and foundations, which is the most commonly used

Figure 11.1 Four pulse-jet filter collectors partially installed, three insulated and jacketed, with air header and diaphragm cleaning valves, structural supports and cross bracing, exhaust outlets, clean-air access, and hopper access doors visible.

strength. Then adequate reinforcing steel and anchor bolts should be used in the concrete to complete a reliable strength foundation. Past experience in the same location or conditions may provide an adequate basis for foundation design, or structural engineers can design it.

Uplift or overturning moments due to wind or seismic loads should be checked so the mass of the foundation can be increased if necessary. Be sure that the specification spells out the available soil strength. Note that the required foundation and concrete strength will be tested during installation; also the reinforcing steel must be supported during the pour so that it stays where it is designed to be for strength.

Structural strength

The supporting steel structure must, of course, be adequately designed for the loading imposed. The structural steel design should be certified for adequate strength for its height and loading. Your company may have a preference for a material other than ASTM A36 steel at 36,000-lb/in^2 strength in structural shapes, or for column sections other than wide flange beams, but those are most commonly used (see Fig. 11.2).

For columns, be sure to consider a square structural steel tubing or a round structural steel pipe section. These have inherently greater column strength than a wide flange, I-beam, or H-beam section since all the steel weight is located out away from the center or neutral axis, and they offer lower-weight structures with the same strength.

Site: 20 lb/ft^2 @ 80-mph wind and seismic zone 1, grade level
Size: 5 ft 2 in × 5 ft 4 in × 33 ft 1½ in overall height: 11 ft 1½ in high clearance

A. Dust collector flat plate area and wind load: 20 lb/ft^2 × 5.33 ft × (33.125 − 11.125) ft = 2345 lb.
B. Lateral shear per column: 2345 lb/4 columns = 584 lb.
C. Weight load: 5266 lb collector + 2000 lb dust + 2800 lb accessories = 10,066 lb total.
D. Weight stabilizing moment = (10,066 lb × 5.17 ft)/2 = 26,021 ft-lb.
E. Wind overturning moment = A × Hcg (height at center of gravity) = 2345 lb × 22.125 ft = 51,884 ft-lb.
F. United Building Code safety allowance = 1.5 × E = 77,825 ft-lb.
G. Excess overturning moment = F − D = 51,804 ft-lb.
H. Uplift on 2 windward columns = G/(2 × 5.17 ft) = 5011 lb.
I. Empty downward load lee side columns = H + (C − 2000)/2 = 9044 lb/column
J. Maximum downward load = (2000/2) lb dust + 9044 lb/column = 11,044 lb/column
K. Column selection = maximum of J or H = 11,044 lb/column at 11.125-ft height. Thus possible columns are
   1. 4 × 4 × ⅜ angle A-36 steel @ 9.8 lb/ft weight, 76,000 lb for two angles, so one column = 76,000/(2 × 1.5) = 25,333 lb. Choice OK.
   2. 3-in^2 structural A-46 steel tubing @ 6.86 lb/ft. 22,000-lb allowable load at 11 ft height and 19,000 lb @ 12 ft, thus more strength and shape stability, plus less weight. Better choice.
   3. Use light angle, 2 × 2 × ¼, cross bracing all four sides, with clearance as needed.

Figure 11.2 Preliminary structural strength calculation for wind, seismic, and weight loading.

Their yield strength is greater at 46,000 lb/in^2 for A500 square tubing and pipe size electric resistance welded, ERW-type steel, and their "closed" shape provides a reduced surface area for painting and lower wind resistance. Although these are minor factors, they can provide worthwhile benefits over the long term and possibly lower initial cost.

Sway bracing is recommended for column heights over 8 ft. Where clearance is needed for truck passage or loading, then columns must be designed strong enough for their unbraced length. The length of each light cross-bracing member is usually limited to 20 ft or less, or 14 ft in vertical height. Beyond that height the usual practice is to use another set of sway bracing members. *A* $2 \times 2 \times \frac{1}{4}$ structural steel angle is generally the lightest bracing member used. The centerlines of the bracing and column members should intersect the top and bottom, or base plate, joint centerlines as closely as possible. This will hold the effect of moments on the connections and columns to a minimum.

Collector installation

With the foundation prepared per plan, and the concrete allowed to cure for 7 days and 90 percent or better strength, the collector installation can begin. Barring any unusual procedures due to project needs, a typical collector installation can proceed as follows. (See Fig. 11.3.)

The first order of business should be to check incoming units for possible damage in shipment upon arrival, and as much as possible before unloading! Any damages should be noted on the bill of lading and signed by the driver. The carrier is responsible for any damage during shipment and should be notified and claims filed immediately. If there is a serious amount of damage, photographs should be taken to back up claims.

Structural steel should be erected first and carefully checked for uniform height, should be plumbed straight vertically, should have correct footprint dimensions, and base and top plates level with correct connection holes. On small to medium collector sizes, the complete collector, or the hopper only, may be shipped with the structure attached. This, or a field assembly, may then be erected as a unit, depending on crane capacity and clearances, to save assembly time up in the air.

A horizontal channel frame or equivalent will normally be used on top of the structure for medium to large collectors, say from 6 ft^2 and up. Then the hopper flange is bolted to the channel frame. If there is a separate dirty-air plenum, this is next joined to the hopper with a sound caulk seal between the mating flanges.

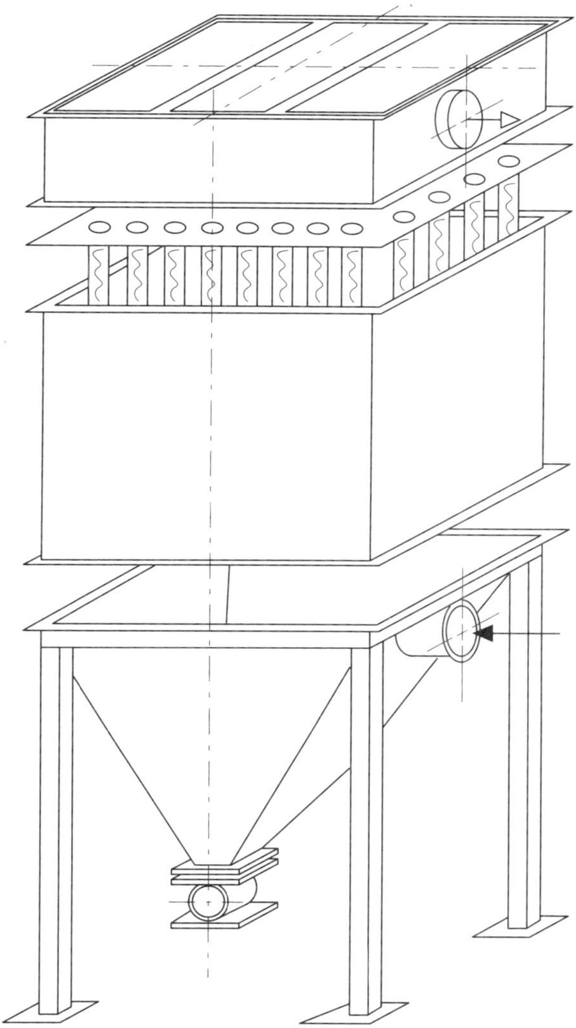

Figure 11.3 Filter collector installation—exploded view, with typical work steps.

Bolt fastener holes will need to be checked and aligned prior to assembly, caulking, and bolting. Since fabrication errors or some twisting or racking in shipment can occur, correcting the problem on the ground is easier than in the air. This alignment should be checked for sure before any joint is caulked for the final seal fastening.

One of the best bolt-up plans for erection is to locate and fasten the center holes first. Then the corner holes can be mated and will have half the variance of an end-to-end difference. This procedure can usually provide for the plenums to be fastened together much more easily. Bolts should be snugged up but not tightened fully until the fit of the entire joint is assured.

The tube sheet, which will hold the filter elements, would normally be attached to the top of the dirty-air plenum during shipment. This helps stiffen both the dirty-air plenum and the tube sheet and protects against handling damage. The tube sheet should be checked for fastener hole fit, lifted in the field enough to allow a good caulk seal to be applied between the tube sheet and dirty-air plenum, and then enough bolts should be fastened for stiffness while the tube sheet is being erected. The fastener fit and seal of the joint between hopper and bottom of the dirty-air plenum should be completed in similar fashion.

Then the clean-air plenum should be erected on top of the tube sheet and dirty-air plenum. This joint will follow a similar check of fastener fit and get a good caulk seal, and the joint should be fastened down securely.

At this point the access platform and ladder will usually be erected. Some manufacturers may provide for an earlier attachment of the access members, or at least attachment brackets. They might also be assembled at the factory or on the ground for smaller collectors, or for top-door access designs. All access provisions must meet OSHA, user, and any other local safety guidelines. While the American OSHA requirements are generally stricter than local or international rules, it is certainly best to check exactly what is required for each installation.

Safety cages on ladders, height of safety railings on platforms, toe board plates, and strength of gratings and other members need to be in compliance. Possible alternative features would be a safety gate at the top of the ladder; intermediate platforms on ladders over 30-ft tall; top or "roof" access ladder and railing, especially for top-door filter access on low-height clean-air plenums; side or hopper access where needed; and sometimes attachment of the ladder to the structural columns to eliminate any attachment to or penetration of the roof of a building.

Other requirements

With the mechanical erection complete, work can progress on all other requirements:

- Utility connections
- Compressed air or gas for filter cleaning

- Air header and blow pipes, diaphragm and solenoid valves
- Electric wiring of the timing control panel and pulse valves as well as any other electrically powered equipment
- Other accessory services such as sprinklers, emission monitors, discharge and handling valves and conveyors, secondary filters, diverter valves and dampers, ductwork, insulation, and heaters if field-installed or completed
- Installation of filters, cages, and any other internal parts, such as grounding connections

Utilities

Electric wiring service must have the correct voltage and amperage capacity and be enclosed and wired to meet the enclosure requirements. Most common would be NEMA standard #4 outdoors, or #12 indoors, and NEMA #7 or 9 in explosion hazard areas. Local code or user requirements must also be met; see Chap. 10 also.

The compressed air or gas supply must have sufficient capacity for the volume consumption and to be able to maintain adequate pressure. This should be considered when planning the installation so that an adequate air or gas supply or piping flow capacity is provided. Consider whether insulated or heated air lines and valve enclosures will be needed in the winter. A pressure gauge is an economical instrument to provide a constant reminder of adequate supply pressure. Provide refrigerated air driers where necessary, and always provide either an automatic or manual moisture drain on the air header at the collector.

Filter media installation

The key to effective dust separation in a filter collector is the filtration medium that captures the dust particles in the airstream. This needs to be emphasized to all installers before the filters are put in place. It should be a simple operation to achieve a leakage-free installation, if carefully planned and carried out!

Snap band tops on filter bags are the simplest and most effective means of sealing the filter in the tube sheet. An accurately formed tube sheet hole is also critical. Punched or laser-cut holes can be held to a close tolerance of 0.015 in, or ±0.0075 in, on a 4.5- or 6.0-in-hole diameter. Other diameters are used and should be of equivalent accuracy. Be sure the hole edges are finished smoothly, of uniform diameter, and without irregularities, which is standard commercial quality. But watch out for poor-quality tube sheet holes during installation or earlier inspection.

A uniformly flat, level, and smooth tube sheet with adequate structural reinforcement is necessary for trouble-free operation, and it helps achieve no leakage around filter element holes. A warped tube sheet surface may create problems with filters not hanging vertically plumb and therefore contacting each other. This can cause abrasion and wear during operation and cleaning. Besides, when the tube sheet projects into the flanged joint between the dirty- and clean-air plenum, any warpage or unevenness may cause joint leakage or a tough sealing problem.

Cages are used inside filter bags to hold the filter element out at the desired diameter since they are usually subjected to a negative or suction pressure in operation. The cage must be smooth and accurately formed to the desired diameter with no sharp points anywhere that might snag the filter. Most felted bags are used with cages that have 10 vertical wires and horizontal wire rings on 8-in centers. More fragile media filters may require up to 20 wires vertically and closer spacing of horizontal rings for better support and less severe flexing during cleaning. Fiberglass, Teflon, ceramic, and coated fabrics or felts may carry a recommendation for these closer construction cages.

The top of the cage will usually have a flat or rolled top flange to support the weight of the cage and bag on the tube sheet. A snap band bag requires a rolled-over flange to protect the bag from damage when workers walk around on the tube sheet. It also supports the flat flange on a perforated diffuser cylinder if used inside the cage and bag. The diffuser provides benefits in better filter cleaning with longer filter life, as discussed before.

Some new collector developments use a catalyst element inside the filter bag for NO_x neutralization which may also need to be supported on the cage flange. Others may be supported on the cage bottom or make use of several horizontal wires supported by a cage ring and may require stronger cages. Another design makes use of a perforated sheet combination cage and diffuser both to support fragile filter media better and diffuse the cleaning air more uniformly along the filter length. This combination also provides better distribution of the cleaned-air flow through the catalyst inside the cage.

Bags, cages, and diffusers, as well as grounding connections, if required, should all be stored in a safe, dry place, such as a warehouse, during the collector installation until needed. These valuable parts can be easily damaged if not protected. (See Fig. 11.4.)

Snap band bags. From the top of the tube sheet, insert the filter bag carefully, bottom first, all the way down into the selected hole. Take special care not to damage the filter surface on the edge of the hole,

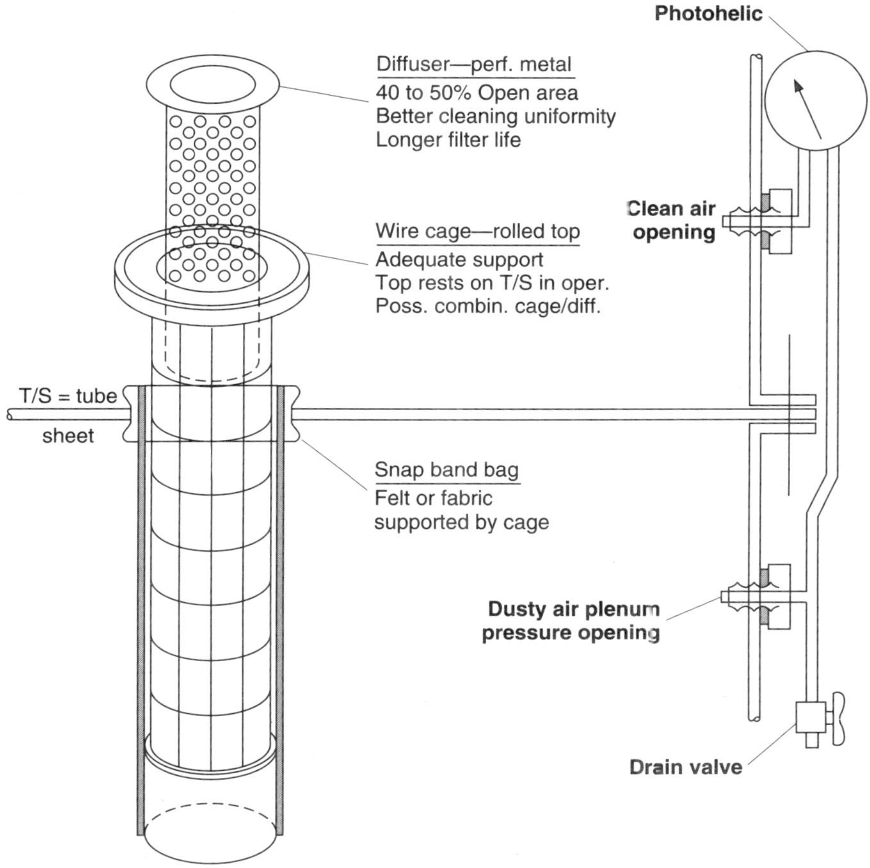

Figure 11.4 Filter, cage, diffuser, and photohelic (or magnahelic), typical installation.

especially if there is a fragile membrane or treatment on the filter surface! Fit the snap band ring partially into the hole so the edge of the tube sheet hole fits snugly into the groove in the center of the snap band. Push a U-shaped bend in the top ring of the snap band inward toward the center, forming a small bend, and grasp the bag by this bend.

Hold the far edge of the snap band against the tube sheet hole, and slowly allow the bend to spring out, while guiding the band so the full circumference of the tube sheet hole fits snugly into the groove in the snap band. When the band springs out into a full circle, there should be a definite "snap" sound.

Visually inspect the band groove in the tube sheet hole to ensure that the fit will have no leaks. Pull upward lightly on the band while running your hand around the inside of the band to make sure the groove is correctly sealed around the full circle. The fit should be snug enough so that the snap band bag cannot easily be rotated in the tube sheet hole.

Cages, and/or diffuser cage assembly. Lower the cage or diffuser-cage assembly into the bag so that the rolled flange of the cage covers the snap band and protects it. The flange edge should rest evenly on the tube sheet surface.

Venturi and clamped bag installation. First install the venturis in the tube sheet holes, per the manufacturer's instructions. A vertical plumb mounting and a tight seal against the tube sheet are the critical first steps.

Fit the cage to the venturi hub, usually with a rolled ring in the cage fitting into a rolled recess in the venturi.

If a diffuser is used, it will usually be inserted into the cage before both are installed on the venturi hub. The cage will have a diffuser support of crossed wires at the correct height in the cage, so the diffuser rests on that support.

Then the fabric filter is brought up from below the tube sheet, slipped up along the cage, and the top of the bag is rolled over at least an inch. This rolled filter cuff is then clamped onto the cage, usually with a stainless-steel clamp. The filter bag must be raised high enough so the bottom of the bag is snug against the cage bottom when clamped, but not pulled so tightly as to cause wear.

An alternative design and procedure allows the bag to be installed on the cage outside the collector. Then the assembly can be brought in and installed at one time. The fabric bag is usually rolled over the top sleeve of the cage before installation and clamping, so the filter bag is supported better, and usually sealed better.

As with any mechanical device, time, unusual service, and wear can create leaks or failure. These should be recorded by location in the collector and by the nature of the failure during the life of the collector (Fig. 11.5). Then any pattern of recurring failure at certain locations, or by type of failure, can be evaluated. Whether it is an unsuspected blast of higher-velocity incoming air impinging on the filters, or sewn seam failures possibly by the cleaning air pressure being too high or due to poor seam fabrication, a solution can be found more easily if the location and cause are known. *Note:* Install all the bags, cages, and/or diffusers very carefully, so the filter elements or bags are not damaged prematurely. They can give good service and last for years with care.

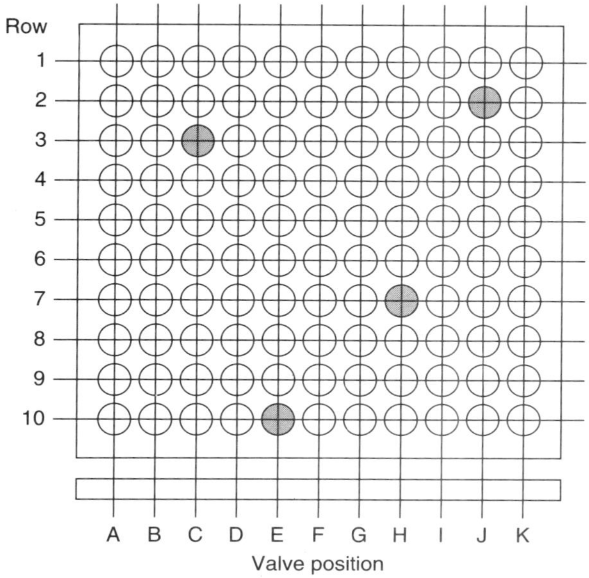

Log of failures:

Valve	Row	Date failed	Date inst'l.	Deterioration or failure	Location of probl.
C	3	12/2/93	1/15/92	Seam	Full lgth.
H	7	12/20/93	1/15/92	Seam	Full lgth.
D	10	12/30/93	1/15/92	Conden.	Overall
J	2	1/5/94	1/15/92	Tear	Poss. instln.

Figure 11.5 Filter element service log and chart.

Magnahelic or photohelic gauge/controller installation

All accessories and a detailed instruction sheet should be packed with the gauge. Mount the gauge or gauge/controller in an easily accessible location for checking during operation and in a place that is free from excessive vibration and accidental damage. It should be where the temperature does not exceed 140°F (60°C) and should be out of direct sunlight. A bracket on one of the structural support columns, along the working platform at the tube sheet level, or another place close to regular collector access, is good.

Connecting the air lines from the gauge to the bag house can be done with ¼-in outside diameter black plastic or copper tubing with compression or quick connect fittings. Pipe thread connections are typically provided on the collector plenum walls, one below and one above the tube sheet level. An in-line filter, a loop, or U-shaped trap should be placed in the tubing from the dirty-side connection, to minimize dust entry into the lines. Then check the tubing lines for leakage and kinks.

Compressed air supply piping

Dirt, scale, or foreign matter can cause blockage in the air lines, valves, or blowpipes. Oil can cause buildup and blinding of the filters. Water in the air can cause valve problems or freeze-up in cold weather. So it is advisable to have an air supply that is clean, dry, and oil-free; and the piping should be blown out during installation to prevent initial fouling problems before connection to critical instruments or valves.

The air receiver or header should have a moisture drain at its low point for best drainage. In-line air filters with automatic drains will suffice if moisture content is not too great and they do not freeze. Insulation and heated lines are advisable in extremely cold climates. A refrigerated drier is recommended in the supply line for outdoor installations if at all feasible.

Piping should be adequate for the airflow demand to minimize pressure drop. Table 11.1 gives general recommendations for pipe supply sizes.

Air consumption should be forecast by the collector manufacturer, and will be based on an estimated cleaning demand, usually at a 0.10-s pulse and a 10-s interval or cycle *off-time*. New controlled air consumption valves and on-demand cleaning will reduce the consumption. For example, with a light dust loading the off-time could be as long as 30 min to several hours, reducing the air requirements from one-third to one-tenth of the original estimate.

The air header and diaphragm valves will usually be installed at the factory. If not, then their mounting instructions should be closely followed. The solenoid valves are usually either mounted as an integral part of the diaphragm valves or supplied in a separate enclosure. This enclosure will usually be shop mounted near the air header but sometimes must be mounted in the field on brackets.

It may be advisable to provide heat tracing, insulation, and even a

TABLE 11.1 Recommended Pipe Supply Sizes

Total free air consumption, ft^3/min (standard)	Piping length		
	0 to 100 ft	100 to 500 ft	500 to 1000 ft
0 to 50	1-in OD*	1¼-in OD	1½-in OD
51 to 100	1¼-in OD	1½-in OD	2-in OD
101 to 200	1½-in OD	2-in OD	2½-in OD

*OD=outside diameter.[1]

heated and insulated enclosure around valves in very cold climates. Past experience at that installation site should be a guide. Consult the manufacturer if in doubt about valves not working in the expected cold climate.

The timer or control panel will usually be shipped loose and then field mounted in a vibration-free location. Ambient temperature should not exceed 120°F (49°C) unless the electric controls are rated for higher temperatures.

Install an on-off or disconnect switch in the line to the control panel, usually a 115-V ac supply. Connect field wiring to the terminals of the solenoid valves as marked.

11.2 Start-Up and Shutdown

Initial start-up

Visually inspect all equipment and remove all foreign material and trash. Check that all guards and safety equipment are installed. Be sure all electric wiring has been checked out.

1. One by one, start up the fan, discharge valves, and conveyors, and any other accessory equipment. Check each one for proper rotation, speed, and operation. Turn each one off and correct problems if necessary; then recheck.
2. Check that all ductwork connections and supports are in place and tight, clean-out ports are closed, every duct is free of debris, hoods are in place and free of contact with machinery, and the fan exhaust is free and clear and aimed correctly.
3. With all equipment turned off, check that all doors and ports are closed on the collector.
4. Turn on the compressed air to the collector cleaning system and check for leaks. If air is leaking from any blowpipe, there may be a leak in the diaphragm to solenoid valve tubing line. Turn off the compressed air when everything checks out.
5. Turn on the electric power to the timer or control panel. The red power ON light should come on. Select the timer position for cleaning, and turn both the off time and the on time knobs to a fully counterclockwise position. The individual timing lights at the output terminals should blink at 1.5-s intervals, and the corresponding solenoid valves should activate audibly.
6. Turn on the air supply to the air header. Set the air supply pressure regulator to about 85 lb/in^2 gauge, unless otherwise instructed. Each diaphragm and solenoid valve should operate every 1.5 s (exhaust air can be felt and heard exhausting from each valve). Let the collector pulse for about 10 min to ensure that all lines are clear. Then set the off time to about 10 s. These settings may need later adjustment in operation.
7. Turn on all dust discharge equipment.

8. If moisture or condensable vapors are present in the airstream, the operating temperature should be brought up above that dew point. It may take some time to get all the ductwork and collector walls above the dew point.

9. Start the fan with the damper set for about half flow volume and run for about 30 min; then start introducing a light load of dust or an intended precoat filter aid powder or sorbent. This is particularly recommended when the process dust will be a very-fine-particle size, and some precoat dust cake is critical.

10. Observe the differential pressure reading on the photohelic or magnahelic gauge. The pressure drop should be very low at start-up. After some 30 min of operation with light incoming dust load, the pressure differential should begin to climb as the bags acquire some coating. The filtration efficiency should then be high enough to begin the feeding step below.

11. The fan damper can be opened up, gradually, and then dust fed at normal design operating rates. The differential pressure should then climb for several hours and stabilize at some value between 1 and 6 in WG. If below 4 in WG, then gradually increase the off-time until it reaches 4 in WG. If it is above 4 in WG then the off-time should be decreased until it settles on 4 in WG. This is a general guide, but use caution on fragile filter media and with very fine dust.

12. Temperature must be controlled to remain below the maximum rating of the filter bags or cartridges.

13. The collector should now be ready for regular operation at the design capacity.

Normal collector start-up

Subsequent start-ups should begin with all systems off (if new filter bags have just been installed, then the initial start-up procedure above should be followed). Normally equipment should be turned on as follows:

1. All ports, doors, and rotating discharge equipment should be closed, and all safety devices should be in place.
2. Turn on the compressed air.
3. After the pressure reaches 85 lb/in^2 or the normal level, turn on the timer.
4. Turn on all dust discharge equipment.
5. Preheat the system if necessary (this may precede the other steps if several hours preheat is required).
6. Turn on the fan and introduce filter aid powder or a low loading of process dust to precoat, then increase dust flow to the normal operating load.

7. If the exhaust is visible, then refer to the troubleshooting guide. A collection efficiency of around ±99.95 percent should eliminate any visible plume unless some steam, vapor or SO_2 is present.

Normal collector shutdown

1. *Dust collection and pneumatic conveying systems.* The shutdown generally reverses the start-up. First turn off the process dust feed, then the fan, unless the two operate in an integrated manner. Wait 5 to 10 min and then turn off the cleaning timer to allow filter cleaning without input dust. Then turn off the compressed air supply, and finally the discharge equipment can be stopped.

2. *Process systems.* Process dryers and equipment feeding the collector should be run until empty of dust, and, if moisture is present, then the heat and fan flow at a reduced rate should be maintained until the collector metal and filter bags are dry. Then proceed as in step 1.

11.3 Safety before Entering Collector Housing

1. Run cleaning system for about 10 min with the fan off to clean the filters well.
2. Run solids out of the hopper discharge equipment during this clean out and inspect for amount of dust and condition.
3. Turn off and lock out all electric power to rotating equipment.
4. If toxic gases are present, purge collector and block off inlet duct or damper, and follow any further safety precautions.
5. Install safety cables and catwalks where needed.
6. Secure access doors in an open position, or remove doors, unless rain or ambient temperature intrusion is undesirable, to prevent trapping anyone inside.
7. Use the buddy system and other safety precautions when people must enter the bag house.
8. Wear respirator if needed in the bag house, especially if carcinogenic dust or toxic vapors are present.
9. Use common sense and any special precautions.

11.4 Troubleshooting Guide

1. High-pressure drop
   - Check filter cleaning system.
   - If cleaning is ineffective, adjust cycle, air pressure, or dust loading, or use off-line cleaning.

- If there is reentrainment of collected dust, check discharge equipment or reduce filtration velocity (A/C ratio).
- If bags are wet, control dew point excursions, dry bags with clean air, clean bags, or replace.
- If the filtration velocity (A/C ratio) is high, reduce airstream volume, or increase filtering area.

2. Low-pressure drop:
   - Check and blow out photohelic gauge lines and protect them from dust or water buildup, or put in a line filter. If tubing is made of copper or other metal, air purge annually or at maintenance shutdowns to remove moisture.
   - May be caused by broken or uncoupled gauge lines.
   - If caused by overcleaning of filters, reduce cleaning pressure or frequency.
3. Dust emission:
   - Check for bag failure on each cleaning pulse, inspect any row that allows emission during cleaning, and check for repeat failures.
   - If there is an increase in bag permeability, check for row and bag, reduce cleaning, and check for bag damage at next shutdown.
   - May be caused by dirty air leaking into clean-air plenum.
   - May be caused by an increase in inlet load.
4. Dust puff back at hoods and inlets:
   - Check for a high pressure drop.
   - If fan volume is low, speed up the drive.
   - The duct pickup system may be out of balance, especially if only some hoods and inlets are a problem.
   - Ducts may be plugged or leaking.
   - Hood design may be poor or damaged, causing poor pickup.
5. Low dust discharge:
   - May be caused by air leaks at discharge.
   - There may be a malfunction of discharge equipment.
   - There may be reentrainment of dust in the collector; check for leaks and heavy dust load.
   - If dust is being retained on bags, increase cleaning or reduce filtration velocity or dust load.
6. Unusual noise:
   - Check it out and correct.
7. Corrosion
   - May be caused by improper paint or application, dew point excursions, improper insulation, emission from nearby equipment, change in process stream, or improper base metal enclosures.

Maintenance

The following maintenance checks should be made regularly:

1. Record daily pressure drop levels.
2. Record filter bag replacement locations and frequency, per Fig. 11.5.
3. Check out unusual bag failures.
4. Maintain a reserve supply of bags and other spare parts.
5. *Daily.* Check cleaning air pressure, drain header, and discharge equipment output; rap hopper with rubber hammer to check for dust buildup or bridging; visually inspect stack plume; check temperature of operation.
6. *Monthly or at maintenance shutdown.* Inspect tube sheet for dust accumulation, corrosion, condensation, bags or cages out of place; inspect filter bags and perform tracer powder leak detection test; check timer for proper operation; check pressure gauge connections and clean if needed; check foundation, housing joints, fan, and motor bolts; check belt and chain drives; check airflow into hoods and collector inlet.

11.5 Testing

For most industrial collection applications, the only air "test" measurements needed can be made with a Pitot tube,[2] for static pressure and total pressure in the ducts and fan, unless an air flow, or volume, measurement is needed. The operating levels of temperature, differential pressure across the collector, compressed air pressure, rotating speed of discharge equipment, and motor load in current or watts will often already be instrumented. For accurate steady-state indication of these operating levels, some type of continuous measurement, recording, and control should be included.

Some form of CEM should be considered since its cost is more reasonable now and the reassurance of knowing that you are in compliance at all times may be worth the cost. With the inclusion of a control relay or an alarm on the CEM, you get instantaneous action when there is an upset or malfunction. See Sec. 7.1, Emission Monitors, for a fuller discussion, latest developments, and costs.

Baseline operating parameters from testing

Input dust loading can be weighed or estimated from process operating rate or discharge output. Exhaust emission dust loading should be read continuously with a dust emission monitor or carefully sampled and correlated for expected operating rates. The operating airstream velocity and volume can be calculated from the static and total pressure test measurements mentioned above. All these measured, calculated, or indicated performance levels should be recorded at initial start-up of normal operation, correlated with acceptable emission or collection efficiency levels, and then used as the baseline operating parameters.

Performance testing

Collection performance testing in operation should include airstream velocity and volume calculated from test pressure measurements, indicated pressures and temperature, input dust loading, discharge equipment dust flow rates, and exhaust emission rate or monitor measurement, or the new programmed emission monitoring software (PEMS). These can then be recorded on a regularly scheduled basis for checking performance.

The frequency should be in accord with regulatory requirements and also for good operating monitoring. Daily visual checks, weekly or monthly performance records, and quarterly for more thorough weight and emission tests would be a reasonable starting cycle. As in any sampling, the frequency should be adjusted in accordance with the variation from standard or normal operating performance.

Pressure measurements

Fans propel the air and airborne dust usually by two methods. When a fan draws air into the fan inlet from the collection system, a negative pressure, suction, or vacuum can be measured in the system. When the fan blows air into a collector from the fan exhaust, then a positive pressure can be measured relative to atmospheric pressure.

Either way, the airstream is moved through the collection system, and a force can be measured in the direction of *flow*. This force or pressure is identified as the *velocity pressure* and the airflow velocity is proportional to V_p. It is usually measured in inches of water column or inches water gauge (see Eq. (9.10)).

In a dust collection system there is a second pressure always present, the static pressure. The total pressure in the airstream is the sum of velocity pressure and static pressure that must be overcome in the system. Static pressure acts equally in all directions.

Exhaust and dust collection systems pull a negative static pressure through the hoods, ducting, and collector into the fan inlet. On the discharge side of the fan and all subsequent ductwork, secondary filters, or exhaust stack, the static pressure will be positive. So the fan is the dividing point, with negative static pressure at the inlet and positive static pressure on the discharge side. The total static pressure that the fan delivers is the sum of the negative and positive static exerted on each side of the fan.

Since total pressure is the combination of static and velocity pressure, it is also usually expressed in the same units of inches water gauge or grams per square centimeter. While velocity pressure is not easy to read directly, it can be determined by subtracting static from total pressure. This subtraction need not be done mathematically, it can be done with a Pitot tube and a calibrated manometer.

A separate static sensor line and a total pressure sensor line can be connected to a manometer. Since both sides of the manometer have static pressure applied to them, the difference or velocity pressure component of total pressure is indicated on the manometer.

A Pitot tube consists of an impact tube with an opening that should be pointed directly into the airstream, or upstream, and thus parallel with the flow. This impact or upstream opening measures total pressure. A second tube of slightly larger diameter is fitted concentrically around the impact tube. The second tube has several radial sensing holes which are thus perpendicular to the airflow and measure static pressure.[3]

As shown in Fig. 11.6, the pair of concentric tubes in a Pitot tube are then bent at 90° and brought out through the duct wall for two separate connections. The inner impact tube connection is brought straight up, and tubing connects it to the high-pressure side of a manometer. The outer static pressure tube is brought out with a 90° attachment and is connected with tubing to the low-pressure side of the manometer.

To be sure that a Pitot tube reads velocity pressure accurately, the total pressure or impact tube tip must be pointed directly upstream, parallel to the airflow. The impact tube tip is exactly parallel to the static pressure outlet connection, so the latter can be used as a pointer outside to align the Pitot tube parallel to the airstream. This can be verified since the manometer pressure will reach a maximum reading when the Pitot tube is aligned correctly, and the impact opening is located in the center of the duct or pipe.

The Pitot tube should be inserted through holes in the duct at least 7½ duct diameters downstream from elbows, bends, or other obstructions, because accurate readings cannot be taken in turbulent airflow. Inclined or long manometers improve the accuracy and can be calibrated to read directly in both pressure and velocity.

The Pitot tube is a primary measuring device and is used to calibrate all other velocity measuring devices. As a standard it should be designed and manufactured to read accurately so that it does not introduce any errors. This allows its use without correction or calibration,

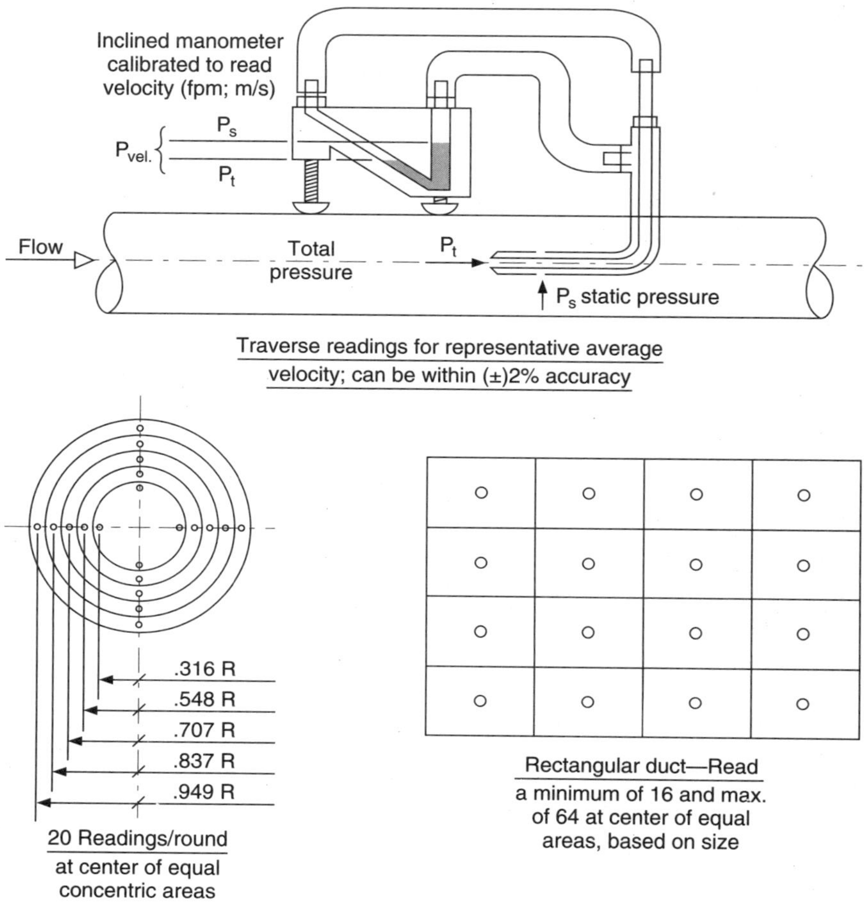

Figure 11.6 Pitot tube and manometer reading pressure and velocity, with traverse readings for representative average velocity in round and square ducts for ±2 percent accuracy.

and it can be so purchased with a unity calibration factor. Pitot tubes are built to AMCA and American Society of Heating and Recirculating Air Engineers (ASHRAE) standards.

In industrial installations, the velocity of an airstream is rarely uniform across the cross section of a duct. Friction slows the air moving close to the walls, so the velocity is greatest in the center of the duct if the measurement point is in a straight duct at least 7½ diameters downstream from fittings in the duct.

To obtain a representative average velocity in ducts with diameters larger than 4 in, a series of velocity pressure readings must be taken at points of equal area in the duct. The approved method is to make two traverses across the duct cross section at right angles to each other. Twenty readings are taken at the center of annular rings of equal area (see Fig. 11.6).

In round ducts at least 20 readings should be taken in two traverses across perpendicular diameters as shown in Fig. 11.6. For round ducts 6 in and smaller, at least six traverse points should be read. For round ducts larger than 6 in diameter, at least 10 traverse points should be read.

In rectangular ducts a minimum of 16 and a maximum of 64 readings should be taken at the centers of equal rectangular areas. Enough readings should be made so the greatest distance between centers is approximately 6 in. Actual velocities for each area are read from the manometer or calculated from the pressure readings. The velocities are then averaged.

By taking Pitot tube readings with extreme accuracy the air velocity can be determined within ±2 percent. For the maximum accuracy the following precautions should be observed:

1. Make the duct diameter at least 30 times the Pitot tube outer diameter.
2. Locate the Pitot tube access holes in the duct at least 7½ duct diameters downstream and 1½ diameters upstream of any elbows, bends, size changes, or other obstructions.
3. Provide an egg-crate type of straightener 5 duct diameters upstream of the Pitot access holes.
4. Make a complete and accurate traverse.

For frequent performance testing between periodic traverse readings, in small ducts, or where traverse readings cannot be made due to obstructions, an accuracy of ±5 percent can usually be achieved by placing the Pitot tube in the center of the duct. Determine the velocity from this one reading and then multiply by 0.9 for an approximate average velocity.

To correct the velocity reading for other than standard air conditions, the actual air density must be known or calculated. The *Industrial Ventilation* manual has a density correction factor table (figure 6.14) that provides correction factors for temperatures to 1000°F and altitudes to 6000 ft.[3]

To determine dry air density, multiply the standard condition density of 0.075 lb/ft^3 by the density correction factor determined above. To correct the air velocity for other than standard air conditions, the actual air density must be known, and can be calculated using the following equation.[2]

$$V_a = 1096.7 \times \sqrt{\frac{h_v}{d}} \tag{11.1}$$

where V_a = velocity of air movement, ft/min
h_v = velocity pressure, in WG
d = actual dry air density, lb/ft^3

Suppliers of Pitot tubes and manometers will also supply an air velocity calculator with each measurement kit that will do the following.

1. Determine a relative humidity factor for the density.
2. Read the density for the barometric pressure and temperature of the location using the relative humidity factor from number 1.
3. Read the actual air velocity in feet per minute, using the density from number 2 and the velocity pressure inches water gauge from the Pitot tube and manometer.

Air volume or airflow rate in cubic feet per minute can be determined from the average velocity and the cross-sectional area of the duct as follows:

$$Q = A \times V \tag{11.2}$$

where Q = airflow volume, ft^3/min
A = duct cross-sectional area, ft^2
V = average air velocity, ft/min, as measured in the duct and corrected in Eq. (11.1)

Other air velocity meters are available and are useful for quick readings. These include hot-wire anemometers, rotating and swinging vane anemometers, velometers, and variable area flow meters. They are often used for low-velocity airflow measurement in hoods or open areas and can be dependably used in correlation with accurate reference flows or total flow in the duct system. All these less-accurate meters should be checked regularly against a primary standard such as a Pitot tube, a high-quality micromanometer, or a hook gauge, or they can be returned to the supplier for calibration.

A carefully planned installation can amplify the results and performance in any collector project. A knowledge of troubleshooting for your specific collector can help you solve simple problems before they cause major downtimes. Performance testing will maximize the results throughout the life of the project. And all these will add to the operating success of your equipment.

References

1. Livingston and Haven Co., Charlotte, N.C., "Engineering Reference," p. 55.
2. Bulletin H-100, Dwyer Instruments Inc., Michigan City, Ind., 1991, pp. 2–5.
3. *Industrial Ventilation,* 1986, pp. 9-1 to 9-10.

Chapter

12

Costs, Savings, Specifications, and Justification

I believe that good design can accomplish pollution control in an enterprise or industry and that companies affected can afford to stay in business!

Twenty-five years ago a good filter dust collector could be estimated and bought for around $1.00 per ft^3/min of filter area. But if you worked at it, you could do better. That was when a reverse-air dust collector and the lowest-cost filter medium were used generally, and the filtration velocity was up around 15 ft/min on sawdust, grain, and other easily collected materials. For many intermittent small-volume operations, an on-off cleaning collector was standard, and for the next tougher applications the step up was a shaker cleaning collector.

Boiler fly ash collection was accomplished with multiple small-diameter cyclones that would withstand operating temperatures of 800 to 1200°F (427 to 649°C) and reduce visible emissions down to the 10 to 20 percent opacity range. It was only on the larger airstream volume continuous process collectors that pulse-jet air cleaning was justified.

Of course, that was an earlier era of pollution control. The limiting guideline then was the amount of visible emission, and emissions were generally acceptable if they were below 20 percent opacity. That was a thin light haze that looked pretty good compared to an uncontrolled full bore plume of dark emissions that deposited dust or fly ash on the neighbors' yards and cars. A small 3-in × 5-in hand-held clear plastic sheet was the typical opacity gauge. It had five levels of opacity, each shaded to represent a 20 percent level of opacity up to black at 100 percent.

Nearby cars were the real telltale gauge, however, because beneath the soot or dust deposit that might settle on them, there usually appeared oily spots that could soon eat through the paint job. No amount of disclaimer could appease irate new car owners; especially when the prevailing wind was in their direction and it happened day after day.

The pollution rules became secondary to the cost of painting all the cars in the neighborhood downwind from the plant; plus the realization that unless this problem were solved it would reappear in about 6 months. The little plastic opacity gauge was no consolation when you stood talking to neighbors about the dust, soot, and rust spots on their new vehicle, especially if their aunts or uncles were on the city council. You had to go back and talk to the boss about doing something more than previous generations had and before the neighbors took legal action. That's when high-efficiency cyclones, wet scrubbers, electrostatic precipitators, better filter media, and pulse-jet filter dust collectors gained respect for their better effectiveness.

Increased investment and operating costs were often justified by how many paint jobs could be saved. The level of emission control that was needed often represented better collection efficiency than the existing pollution regulations. Today, the pollution regulations and emission limits usually determine the collection efficiency required.

12.1 Costs

I believe that good design can accomplish pollution control in an enterprise or industry and that these companies can afford to stay in business! Emission limits must obviously be met in order to improve the environment, but that justification for each individual investment can be some return on that investment through savings and profits. So look for possible savings and profits on any pollution control project, while still meeting the regulations, and keep employment up while we clean up the environment.

The following tabulated costs are estimated on the basis of 1993–1994 cost levels. Cost indices for future years can be used to factor these estimates for the appropriate year. The impact and development of lower-cost designs or high-temperature components, for instance, cannot be foreseen with any accuracy, so these would have to be determined for each individual case. See Fig. 12.1 for a chart comparison of cost for three types of filter collectors.

12.2 Cartridge Filter Costs

Total installed costs for cartridge-type filter collectors may range from \$1.50 to \$8.50 per ft^3/min of process air volume capacity, not including ductwork or utilities, as shown in Table 12.1. Note the

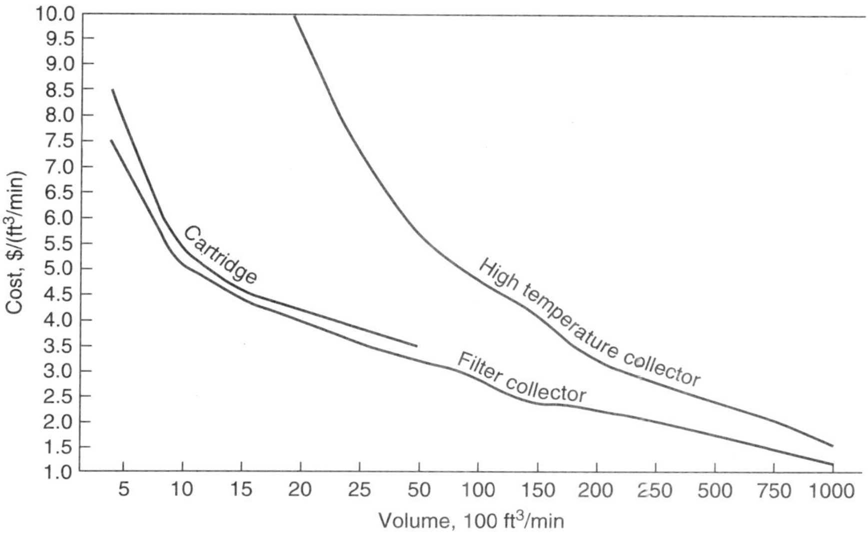

Figure 12.1 Approximate cost versus airstream volume for fabric, high-temperature, and cartridge filter collectors, 1993 cost basis.

advantage in height of cartridge units over bag designs; this is especially important for inside or restricted height locations. Another advantage is the ease of replacing cartridges, and most designs make this possible from outside the collector housing, because of the shorter-height filter elements and their rigid construction.

More and more improvements are being made available in cartridge collectors to bring their operating performance up to the level of bag filter collectors. For these special design and operating improvements a cost increase can be expected of 2 to 4 times the standard cost shown in Table 12.1. These cost increases will vary in proportion to the number of special features used such as:

- Wider cartridge pleats for collection of fines and difficult dust products, and thus less filter area
- Filter media materials, membranes, and temperature of operation
- Corrosion-resistant materials of construction
- Emission limit levels required and the use of secondary filters
- Many accessory additions

12.3 Bag Filter Costs

The total installed cost for standard fabric bag filter collectors usually varies from a low of $1.25 per ft^3/min on very large normal or stan-

TABLE 12.1 Approximate Costs of Cartridge Filter Collectors*

Number of cartridges	Filter area, ft^2	Volume, ft^3/min @ 8 in SP†	Fan, hp	Cost (collector and fan), $	Ratio of $:$ft^3$/min	OA dimensions‡ (W × D × H)§
2	480	960	3	8,100	8.44	3 × 3 × 10
4	960	1,920	7.5	10,750	5.60	5.5 × 3 × 10
6	1,440	2,880	10	13,300	4.62	4 × 5 × 11
8	1,920	3,840	15	15,900	4.14	5.5 × 5 × 11
10	2,400	4,800	20	18,500	3.86	7 × 5 × 11
10	3,000	4,500¶	15	16,500	3.62	7 × 4.5 × 11
12	3,600	5,400¶	15	17,750	3.27	9 × 4.5 × 11
16	4,800	7,200¶	20	22,500	3.10	12 × 4.5 × 11
120	28,800	45,000¶	150	71,200	1.59	30 × 9 × 18
138	33,120	51,750¶	150	76,500	1.48	34.5 × 9 × 18
156	37,440	56,160¶	200	88,000	1.56	39 × 9.5 × 19
180	43,200	64,800¶	200	98,000	1.52	45 × 9.5 × 20
208	49,920	74,880¶	250	112,750	1.51	39 × 12 × 21
240	57,600	86,400¶	300	130,500	1.51	45 × 12 × 22

*Service temperature=180°F, V_f=2.0 ft/min maximum, cartridge filter housing is made of carbon steel, and a polyester felt media is assumed.
†SP=static pressure.
‡OA=overall, or outside, dimensions.
§W=width, D=depth, H=height.
¶V_f=1.5 ft/min.

dard design collectors of around 100,000 ft^3/min airstream volumes, up to around $6.00 per ft^3/min on small standard collectors of around 1000 ft^3/min volumes not including ductwork or utilities. See the 20 various-size collectors and their costs outlined in Table 12.2.

The cost range will increase for large-airstream-volume special designs to around $2.00 per ft^3/min, and for high temperatures to $12.00 per ft^3/min. The actual cost level will vary in proportion to the number of features added, as follows:

- Temperatures above 300°F (149°C) and up to 1000°F (538°C)
- Corrosive airstream exposure that requires special metals such as stainless-steel or special coatings on the enclosures
- Extremely fine particle sizes that require more expensive filter media or lower filtration velocities
- Added requirements such as heavier wall construction for abrasion resistance, food grade, or especially smooth surface finishes; higher design pressure, wind loads, or seismic zone locations requiring increased structural strength
- Other accessories, such as automatic fan dampers, complex control panels, level controls, CEMs, vibratory bin discharge outlets, hop-

TABLE 12.2 Approximate Costs—Fabric Filter Collectors

Number of bags/length, ft[a]	Filter area, ft^2	Airstream volume, ft^3/min	Filtration velocity, ft/min	Fan, hp[b]	Cost, $ per ft^3/min[c]	Overall dimensions, (L × W × H) or diam. × ht.
Ambient to 300°F (149°C) Temperatures						
36/8/B	442	2,000	4.53	10	3.63	4.5 × 5.3 × 10
49/6/B	453	2,265	5.0	10	4.78	5.2 × 6.2 × 12.5
25/8.6/A	332	2,362	7.1	7.5	4.43	3.7 × 3.7 × 24
49/8/AD	602	2,400	3.98	10	5.54	6.3 × 5.5 × 17.3
42/10/A	645	4,000	6.19	15	4.05	5.4 × 5.5 × 29
81/12/A	1,495	9,000	6.02	40	3.42	7 × 8.1 × 35
144/8/AD	1,771	10,000	5.64	40	3.02	7 × 12.5 × 19
120/8/AD	1,162	10,000	8.6	40	2.70	7 × 11.5 × 19
208/12/A	3,839	12,500	3.25	50	3.14	11.5 × 12.3 × 36
132/12/A	2,436	12,800	5.25	50	2.81	9.3 × 9.8 × 37
192/10/A	2,953	16,500	5.6	60	2.47	10.7 × 12.3 × 35
228/10/B	3,506	17,500	4.99	60	2.70	16.8 × 9.3 × 20.5
396/12/A	7,276	29,000	3.99	100	2.75	29 × 9.3 × 37
304/12/A	5,610	32,000	5.7	125	2.25	17 × 12.3 × 38
540/12/A	9,965	39,850	4.0	150	2.68	39.7 × 9.3 × 36
630/12/A	11,625	46,495	4.0	150	2.42	37.0 × 11.5 × 35
672/12/A	12,403	50,000	4.03	200	2.91	32 × 12.3 × 45
630/12/A	11,625	58,150	5.0	200	2.02	37 × 11.5 × 35
810/10/B	12,460	92,930	7.45	300	1.00	39.5 × 13.8 × 20 .5
1,072/14/A	23,082	115,450	5.0	400	1.57	51 × 13.3 × 45
Special Applications for Temperature, Accessories, and Materials of Construction						
S-1 42/10/A[e]	645	2,000	3.09	10	9.48	5.3 × 5.5 × 31
S-2 40/3/C[f]	185	672	3.63	3	84.7	5 diameter × 12.5 ht.
S-3 49/10/A[g]	753	3,000	3.98	10	8.78	6.3 × 5.5 × 21
S-4 137/7/CD[g]	1,475	5,458	3.70	20	10.23	10 diameter × 28 ht.

[a]A=hopper bottom, B=bin vent bottom, C=round design, D=top doors for low-height clean-air plenum.
[b]Fan horsepower as estimated for a typical 10-in WG static pressure.
[c]Ratio of estimated total cost divided by normal airstream volume in cubic feet per minute.
[d]Overall dimensions for a typical filter collector with a reasonable can velocity V_c of 200 ft/min or less. L=length, W=width, H=height.
[e]Less lime injection system @ $5,500+fan silencer @ $2,500 reduces cost per ft^3/min to $5.48.
[f]Includes titanium grade 2 construction and 316L stainless steel accessories.
[g]Includes 304L stainless steel construction and accessories.

per or full housing heaters, insulation, explosion vents, sprinklers, flame suppression systems, fan silencers, and filter aid, lime, or other sorbent injection systems

Unusually severe requirements for corrosion resistance or for a multitude of accessories can boost these costs up to a range of $6.00 to $85.00 per ft^3/min. See S-1 to S-4 in Table 12.2 for the following:

- Corrosion-resistant construction using Inconel, Monel, or titanium grade 2 construction.
- More stringent absorbent features for SO_2 or NO_x emission control.
- Operating temperatures above 500°F (260°C) are feasible with metal or ceramic filters and high-temperature accessories; and the costs for these higher-temperature collectors should come down as quantity production of components increases and solutions are developed for lower-cost construction.
- A large number of accessories on one collector.

12.4 Approximate Fabric Filter Media Costs

These costs are approximate average costs for the most used filter media. A 16-oz polyester felt is the most common for temperatures up to 275°F (135°C), followed by polypropylene felt for 200°F (93°C), and then P-84 felt, Nomex felt, fiberglass fabric and Teflon for temperatures up to 375, 425, and 500°F, respectively (190, 218, and 260°C), and the newest ceramic and metal filters for operation to 1200°F (649°C), and above. These are typical costs for a 6-in-diameter, 8-ft-long filter bag with a snap band positive top seal (except ceramic and metal filter bags are clamped), with a filter area of 12.3 ft^2 per bag. (See Table 12.3.)

Table 12.4 shows the additional costs for surface coatings of polyester Surfcoat, antistatic, Teflon, and Goretex or Tetratex.

A ceiling of 500°F (260°C) has been the norm for industrial filter operation up until recently, which requires the use of a heat exchanger upstream of filter collectors to reduce any higher process temperature below 500°F or air dilution with cooler air. Current developments of higher-temperature filter media are providing test operations up to 1200°F, with potential for higher temperature. So the use of ceramic,

TABLE 12.3 Approximate Fabric Filter Media Costs

Material	Cost per bag, $	Cost per foot, $	Maximum temperature, °F	°C
Polyester	8.50	1.07	275	135
Polypropylene	8.00	1.00	200	93
P-84	16.00	2.00	375	190
Nomex	22.00	2.75	425	218
Teflon	26.00	3.25	500	260
Fiberglass	24.00	3.00	500	260
Ceramic	48.00	6.00	1200	649
Metal	56.00	7.00	1400	760

TABLE 12.4 Approximate Costs of Filter Membranes

Material	Cost per 8-ft-long bag, $
Polyester Surfcoat	1.00
Antistatic	1.00
Teflon B	1.30
Teflon porometric Goretex or Tetratex	2.00

porous metal, or other suitable filter media will offer a straight-through handling of high-temperature process gases in their acceptable ranges, without the use of heat exchangers or dilution.

The work being done on a SO_x-NO_x-Rox-Box (SNRB) process collector is a prime example of higher-temperature possibilities in a filter collector. The three main components of collection are:

1. Ceramic cloth tubular filters that should operate at temperatures to 2000°F (1093°C).
2. Lime or ammonia neutralization of SO_2 by injection ahead of the filter. This provides reaction time in the airstream and during residence time in the filter cake on the bags.
3. Catalyst, held inside the cage-diffuser and bag that receives air cleaned by the SO_2 sorbent and filtration, which allows a more effective removal of NO_x.

This system has proven in preliminary tests that it can successfully remove each contaminant in compliance with emission requirements, at an optimum operating temperature of 800 to 850°F (427 to 454°C). This temperature should also be suitable for most accessories using commercially available high-temperature gaskets, bearings, and construction materials. This development should make the highest collection efficiency available for higher-temperature applications.

12.5 Specifications

The scope of specifications for a filter dust collector can vary widely, depending on the application and process requirements. But they should

1. Specify the process parameters and dust loading requirements
2. Detail emission limits, materials of construction, discharge and accessory equipment preferences, site conditions, structural loading, performance testing, and any other requirements

3. Spell out delivery, financial, start-up, and other factors to be met

12.6 Typical Specification and Format

1. Scope
   a. This specification covers the requirement for a pulse-jet-type negative-pressure filter dust collector, exhaust fan, discharge screw conveyor, and double tipping valve to remove dust particulate from a cement clinker cooler. The supplier is free to propose alternative arrangements and designs in order to achieve a more efficient or cost-effective result.
   b. This specification should include
      (1) Collector housing complete from inlet to outlet flange
      (2) Structural supports, ladder or stairs, walkway or platform, handrails and access safety protection, and a crane or hoist if required for maintenance handling
      (3) Dust collection filter bags and bag support hardware, plus a 5 percent spare part inventory
      (4) Filter cleaning system including ASME code rated air receiver, pulse valves, timer and controls, header and pulse blowpipes
      (5) Collector to include dust hopper, dirty-air and clean-air plenum housing
      (6) Exhaust fan with inlet damper, V-belt drive, guard, and motor or outlet damper
      (7) Electric controls as specified
      (8) Expansion joints, gaskets, caulk, and sealants for leak-free operation
      (9) Shop painting
      (10) Dust discharge screw conveyor and double tipping valve after the screw
      (11) Performance testing and reports including the use of necessary test equipment, transportation, and expenses (tests to be performed by an independent laboratory or service)
   c. This specification should not include
      (1) State, local, or federal taxes
      (2) Building or emission permits
      (3) Equipment installation and erection
      (4) Ductwork to and from the collector and fan inlet and outlet flanges, except for the fan outlet Evasé stack
      (5) Foundations or anchor bolts, except for design loading requirements
      (6) Electric wiring, lighting, or grounding
      (7) Compressed air supply and piping

(8) Finished coat of paint

2. Process requirements. General description. The dust collector is to clean 384 tons/day of process dust from the airstream.
   a. Plant location.
   b. Altitude above sea level is 300 ft.
   c. Ambient temperature range is +10 to +100°F (−12 to 38°C).
   d. Airstream volume is 10,000 ft^3/min (actual) at 250°F (121°C).
   e. Estimated inlet grain loading is 5 to 10 gr/ft^3 (actual).
   f. Maximum allowable outlet grain loading is 0.01 gr/ft^3 (actual).
   g. Normal production is 16 tons/h=384 tons/day=534 lb/min.
   h. Product is cement clinker dust at 60 lb/ft^3 density.
   i. Operating time is 24 h/day, 335 days/year.
3. Performance testing and guarantee
   a. The dust collector shall remove particulate dust from the airstream down to 0.010 gr/ft^3 (actual) or less at the outlet of the collector; based on a maximum input loading of 10 gr/ft^3 (actual) for 99.9 percent efficiency, or 100 gr/ft^3 (actual) for 99.99 percent efficiency; per agreed guarantee conditions.
   b. Performance tests shall be in accordance with EPA regulations and test procedures per #40 CFR, Chap. 1, part 60, Methods 1, 2, 3, and 5 as revised to date.[1]
      (1) *Method 1.* Sample and velocity traverses for stationary sources.
      (2) *Method 2.* Determination of stack gas velocity and volumetric flow rate.
      (3) *Method 3.* Gas analysis for carbon dioxide, oxygen, and dry molecular weight.
      (4) *Method 5.* Determination of particulate emissions from stationary sources.
   c. Performance test report shall specify results below:
      (1) Normal air flow:______ft^3/min (actual) at______°F, out of exhaust stack (Method 2)
      (2) Normal dust input:_______lb of particulate per hour (Method 5)
      (3) Normal dust input loading:______gr/ft^3 (actual) at______°F (Method 5)
      (4) Normal dust emissions:______gr/ft^3 (actual) at______°F (Method 5)
      (5) Normal dust emissions:______gr/DSCF (dry standard and cubic foot) at ______°F (Method 5)
      (6) Upset airflow:______ft^3/min (actual) at______°F, out of exhaust stack (Method 2)
      (7) Upset dust input loading:______gr/ft^3 (actual) at______°F (Method 5)

(8) Upset dust emissions:______gr/ft^3 (actual) at______°F (Method 5)

(9) Normal differential pressure:______in WG at______°F across collector.

(10) Upset differential pressure:______in WG at______°F across collector.

(11) Normal velocity (average of Pitot traverse):______ft/min at______°F at inlet to collector (Method 1)

(12) Normal velocity (average of Pitot traverse):______ft/min at______°F at outlet of collector (Method 1)

(13) Upset velocity (average of Pitot traverse):______ft/min at______°F at inlet to collector (Method 1)

(14) Upset velocity (average of Pitot traverse):______ft/min at______°F at outlet of collector (Method 1)

(15) Normal gas analysis at______°F:______flow ft^3/min (actual) for carbon dioxide, oxygen, and dry molecular weight (Method 3)

(16) Upset gas analysis at______°F:______flow ft^3/min (actual) for carbon dioxide, oxygen, and dry molecular weight (Method 3)

(17) Efficiency of particulate collection:
   (*a*) Normal conditions efficiency = [(dust load input − dust load output)/(dust load input)] × 100.
   (*b*) Upset conditions efficiency = [(dust load input − dust load output)/(dust load input)] × 100.

(18) Test equipment used and measured or calibrated accuracy. The *Industrial Ventilation* manual published by the American Conference of Governmental Industrial Hygienists covers the “Testing of Ventilation Systems” rather well for airflow in ft^3/min, velocity, and pressure.[2]

The Pitot tube is the standard method of measuring air velocity; and by calculation using Eq. (11.2) the quantity of airflow Q can be determined after measuring the cross-sectional area of the appropriate duct. For accurate measurement of average velocity the Pitot tube is used in a traverse method to measure the velocity pressure in a number of equal areas of the duct, since the velocity is not uniform across the duct area. Methods are outlined in *Industrial Ventilation* for both round and rectangular duct traverses that produce a consistently representative set of velocities that can be averaged for an accurate average velocity of flow, as discussed in Chap. 11 and Fig. 11.6.

$$Q = A \times V \tag{11.2}$$

where Q = quantity of air flowing, ft³/min (m³/min)
A = cross-sectional area of duct or hood at the measurement location, ft (m²)
V = average linear velocity, ft/min (m/min)

(19) Accuracy of test method and equipment
  (*a*) Maximum accuracy required for EPA pollution test reporting and recording, calibrated.
  (*b*) Equipment performance accuracy for acceptance or corrective action by supplier, measured.
  (*c*) Operating accuracy for regular performance test and tune-up by plant, measured.

d. Supplier shall furnish capable start-up specialists as necessary and agreed upon to meet the performance guarantee.

e. Supplier is responsible for the expenses of performance testing by a certified independent test lab or service including the use of test equipment, transportation, and reports.
  (1) If the performance guarantee is met, the supplier shall have no further obligation for testing.
  (2) If the performance guarantee is not met, the supplier will be given reasonable time and access to make improvements, after which the testing will be repeated at the cost and expense of the supplier until the performance is met.

4. Design criteria
  a. Dust collector
    (1) Dust collector shall be designed for outdoor service and be of the pulse-jet type. Filter bag removal shall be from the top of the tube sheet through a walk-in plenum.
    (2) All equipment shall comply with current OSHA standards.
    (3) Design negative pressure shall be 20 in WG
    (4) Density of the clinker dust shall be 60 lb/ft³ for volume and 85 lb/ft³ for structural calculations.
    (5) Maximum pressure drop across the collector, flange to flange, shall not exceed 5 in WG.
    (6) Filter bag length may be 10 ft maximum (3.05 m).
    (7) Filtration velocity (A/C ratio) shall be 5.5 ft/min maximum at normal operation (1.68 m/min).
    (8) Inlet duct into the hopper shall be enlarged through a 30° included angle transition to reduce airstream velocity to or below 2000 ft/min exiting into the hopper (610 m/min).
    (9) Hopper slope shall be 55° minimum in the valley angles,

60° minimum on the walls, and any short internal projections shall slope at least 45° from the horizontal.

(10) Inlet and outlet connections shall have $\frac{3}{16}$-in-thick flanges.

(11) Filter cloth area used in filtration velocity calculations shall be only the active area and not include cuff, caps, or other covered areas.

(12) Exterior of the collector shall have an SP-10 blast cleaning and one coat of suitable high-temperature primer paint at least 2 mils DFT (dry film thickness).

b. Structural design

(1) Structure shall be designed in accordance with applicable building codes.

(2) Structural supports shall be of rigid-frame design without diagonal or cross bracing, and allow 4-ft 0-in vertical clearance below the double tipping valve outlet.

(3) All walkways shall be designed for 100-lb/ft^2 live loading.

(4) All walkway bar grating shall be galvanized steel with 100-lb/ft^2 load rating.

(5) All stairs shall be 36 in wide and walkways 48 in wide.

(6) Structural steel shall be ASTM A-36 or stronger.

c. Housing design

(1) Housing shall be a minimum of $\frac{3}{16}$-in-thick A-36.

(2) Reinforcing stiffeners shall be located on the exterior and provide a uniform surface for installing insulation if needed.

(3) Roof of the clean-air plenum shall be sloped at $\frac{1}{4}$ in/ft for water drainage.

(4) Two access doors to the clean-air bag removal plenum shall be 30 in wide × 48 in high, and two hopper access doors shall be 24 in × 24 in minimum.

d. Hopper

(1) Trough-type hopper shall be provided with maintenance access for conveyor hanger bearing locations from the outside.

(2) Dust holding capacity of the hopper shall be 6 h at 10 gr/ft^3 (actual), or calculated holding volume up to the centerline of the airstream inlet.

e. Tube sheet

(1) Tube sheet shall be $\frac{1}{4}$-in-thick A-36 steel.

(2) Filter bag spacing shall be such that the can or upward passage velocity between the bags does not exceed 225 ft/min (68.5 m/min).

(3) Snap-in or snap-band-type bag attachment is preferred, with top bag removal.

f. Filter bags and bag cleaning
  (1) Filter bags shall be 14-oz Nomex for a maximum service temperature of 375°F (190°C).
  (2) Alternative filter materials may be quoted for similar ratings and performance.
  (3) Supplier shall quote the estimated compressed air consumption for bag cleaning in normal service.

g. Instrumentation
  (1) Supplier shall furnish all controls required to operate the collector and shall describe them in the proposal.
  (2) Instrumentation shall include local gauges or monitors and provide remote indication of differential pressure, temperature, and low and high hopper levels.
  (3) The cleaning shall be both sequentially timed or on-demand, selected by a switch on the panel, and adjustable set point differential pressure levels supplied on a photohelic gauge/controller. A remote 4- to 20-mA output signal shall be provided for monitoring or PLC input. The PLC will be supplied by others.
  (4) The control panel shall have a NEMA #4 enclosure and be completely wired and tested.

5. Discharge equipment

a. The screw conveyor shall discharge from the full hopper trough length and extend 2 ft beyond the support structure for outlet discharge.
  (1) Conveyor Equipment Manufacturer's Association (CEMA) standards shall be used for determining reasonable design conveyor speed, capacity, degree of trough loading, and drive power for a 40-rpm maximum rotating speed.
  (2) Conveyor shall operate for 24 h/day and 7 days/week.
  (3) Conveyor shall be weather-tight and airtight for up to 10-in WG vacuum operation.
  (4) Conveyor shall be designed for reversible operation.
  (5) To determine the conveyor component sizes, 240 percent of the motor full load torque shall be used.
  (6) No hanger bearings shall be located under the feed points.
  (7) Drive power shall be adequate for 100 percent cross-sectional loading at start-up.
  (8) The pipe shaft shall have at least a 4-in diameter and be approximately one-third of the screw flight diameter, using Schedule 80 pipe, per American Standards Association specifications.
  (9) Flights shall be right-hand sectional design.
  (10) Flights shall have be a minimum ½ in thick.

(11) Screw sections shall be straightened for a maximum allowable run out of ⅛ in.
(12) Hanger bearings shall be hard iron in type 316 expansion hangers. Tail bearings shall be external antifriction sealed bearings.
(13) Each hanger bearing shall have an access cover adequate to expose all coupling and hanger bolts.
(14) The trough shall be a U-type of $\frac{3}{16}$-in-thick steel plate minimum. Discharge shall be a plain type flanged to accept the double tipping valve or RAL.
(15) Drive shall have a shaft mounted reducer adequate for start-up torque with a 100 percent fully loaded cross section.
(16) Motors shall be TEFC (totally enclosed fan cooled) 1750 rpm 3-60-480 V ac (three-phase, 60 Hz, 480 V ac) rated, with cast-iron housings.
(17) A motion detector switch shall be provided on the tail end shaft.

b. Air seal discharge valve
(1) An air seal valve of RAL design shall be provided to allow no more than 15.0-ft^3/min leakage into the dust collector through the discharge valve at normal differential pressure of 12 in WG negative and temperature of 300°F.
(2) The design operating speed shall be a maximum of 15.0 rpm for continuous operation with the normal dust load and a pocket fill factor of 75 percent maximum.
(3) The materials of construction shall be suitable for long-term use with the dust product specified with its inherent abrasiveness and corrosiveness, and guaranteed for a minimum of 1 year's service at the design conditions specified.
(4) The rotor vanes shall be adjustable for wear so the clearance and leakage rate can be maintained by adjustment no more often than every 3 months.
(5) The ball bearings for shaft support shall be located outboard of the valve housing on each end of the shaft, and an air purge shaft seal installed inboard of each bearing.
(6) The motor and drive on the valve shall be adequate to turn the rotor at the design loading and speed in continuous service 24 h/day; with a shear pin, motor overload, or thermal cutout to protect the valve from damage if an overload or jam should occur.
(7) There must be a slow speed sensor or switch with a 5-A contact for 115-V ac control hookup to provide action or an

interlock in case the rotor tail shaft fails to turn after start-up.

(8) Alternatively, there may be a double dump or double tipping valve used as the discharge air seal valve in place of the RAL. The same leakage, zero speed switch, materials of construction, and 3-month minimum adjustment requirement to maintain design operation, and overload protection must be provided, but the normal maximum operating speed will be 6 cycles/min.

6. Fan
   a. The exhaust fan shall have sufficient capacity in volume and static pressure delivery for the specified system operation.
      (1) The fan shall be designed for industrial continuous duty at the 300°F (149°C) airstream temperature with a heat slinger to circulate air over the shaft and inboard bearing.
      (2) The fan should preferably be a backward-inclined blade type, with a drain and access door in the housing.
      (3) The fan shall have an automatically controlled inlet damper, with a motor actuator.
      (4) The supplier may quote a recommended alternative fan design type, direct drive or variable-speed controlled for efficiency or energy savings.
      (5) The evasé fan exhaust stack shall be quoted as an alternative with power required if different.
      (6) The noise level at 3 ft from the fan while in normal operation shall be quoted, but shall not exceed 85 dba (adjusted decibels). A silencer should be quoted if necessary to reach this noise level (AMCA bulletins 301 and 302).
7. Plant layout
   a. A plant site layout is attached showing the suggested location with space available for the collector, and with preliminary ductwork and fan exhaust stack locations shown.
   b. The supplier shall supply a preliminary layout to scale with the proposal showing a plan and two elevation views of the proposed collector, fan, and stack with overall dimensions shown.

12.7 Cost Justification

If the only justification there is will be to meet the emission limits, then the project will be a straight cost-of-doing-business consideration. If there are estimated cost savings, then the normal justification method used by your company will apply. *If* you need estimates as the project proceeds or there is not a company justification method in place, *do* use the *return on investment* method, which simply put is

$$\text{Return} = \left(\frac{\text{savings}}{\text{investment}}\right) \times 100 \qquad (12.1)$$

$$= \left(\frac{\$15{,}000}{\$35{,}000}\right) \times 100$$

$$= 42.8 \text{ percent}$$

which is good for pollution control and usually takes some real digging to develop. But use only bonafide savings so that they can be confirmed after start-up and after 1 year's operation.

12.8 Conclusion

The purpose of this book is to provide a concise guide for developing the optimum filter dust collector for each project or emission requirement. The goal was also to be objective in evaluation, yet comprehensive in scope. It is intended to be especially useful for developing estimated solutions when time before a deadline is short and where the emission limits are tight. It should also serve as a guide for ongoing performance testing, maintenance, and troubleshooting. The aim is to accomplish, as well as maintain, the optimum performance, efficiency, and cost levels.

Some new technology has been discussed for potential consideration and use, as follows:

- High-temperature filter media for up to 1200°F (649°C).
- Combined function filter collectors for SO_2, NO, and NO_2, and dust collection operation at 850°F (454°C) temperature.
- Recommendations for designing optimum airstream velocities at critical points in the collection system.
- Case histories of some typical dust collection problems and their solution.
- Reduced consumption compressed air valves and systems for pulse-jet cleaning of filters.
- Energy reduction possibilities on hoods and inlets, ductwork, fans, discharge feeders, precoating filters, and other areas.
- Comparison of emission monitors.

- Estimated costs for collectors and pertinent accessory equipment have been provided for reference and as guides, on a 1993–1994 basis.
- Throughout, the intent is also to provide guidelines for the specification and selection of the optimum filter dust collection system available, for the emission limits and operation required, and at the best cost-effectiveness.
- Optimizing filter collector system design and sizing for the best performance and cost.

Filter dust collectors usually fill the need better than any other industrial method for high-efficiency collection of fine dust particulate at the best cost-effectiveness. They can also maintain a highly dependable level of operation over a long period of time, and they deliver an easily reclaimable dry dust discharge.

If this book provides a guide toward achieving the highest separation efficiency available for dust collectors and their reduction of air pollution at the lowest cost, it will have accomplished its purpose. If it sparks optimum cost-saving results, it will have further achieved a goal similar to Chuck Yeager's in aviation, "You've got to know when to push it, and when to back off!"

References

1. EPA regulations, No. 40 CFR, Chap. 1, pt. 60, Methods 1, 2, 3, and 5 as revised to date.
2. *Industrial Ventilation,* American Conference of Governmental Hygienists, 1986, pp. 9-1 to 9-31.

Conversion of Units

Length

inch, in = 25,400 μ
= 25.4 mm
= 2.54 cm

foot, ft = 0.3048 m
= 12.0 in

mile, mi = 5280 ft
= 1.6093 km

micron (micrometer), μ
= 0.00003937 in

millimeter, mm = 0.03937 in
= 1000 μ

centimeter, cm = 0.3937 in

meter, m = 3.2808 ft
= 39.37 in

kilometer, km = 3280.8 ft
= 0.6214 mi

Weight

ounce, oz = 28.3495 g
= 437.5 gr

pound, lb = 453.6 g
= 7000 gr

grain, gr = 0.0001429 lb

grain/ft^3 = 2.288 g/m^3

0.01 gr/ft^3 = 0.02288 g/m^3

gram, g = 0.0353 oz
= 15.4322 gr

kilogram, kg = 2.2046 lb
= 15,432.2 gr

grain, gr = 0.0648 g

g/m^3 = 0.437 gr/ft^3

0.01 g/m^3 = 0.00437 gr/ft^3

Volume

cubic inch, in^3 = 16.3872 cm^3

cubic foot, ft^3 = 1728 in^3
= 28,317 cm^3
= 0.028317 m^3
= 7.48 gal
= 28.3162 L

gallon, gal = 3.7853 L

ft^3/min (actual) = 0.028317 m^3/min
= 1.699 m^3/h

cubic cm, cm^3 = 0.06102 in^3

cubic meter, m^3 = 1,000,000 m^3
= 61,023 in^3
= 35.3145 ft^3
= 264.1562 gal
= 1000 L

liter, L = 0.0353 ft^3

m^3/h = 35.315 ft^3/h
= 0.5886 ft^3/min

Velocity

ft/min = 0.3048 m/min
= 18.288 m/h

m/min = 3.2808 ft/min

m/h = 0.05468 ft/min

Area

Square foot, ft^2 = 0.0929 m^2
= 144 in^2

Square meter, m^2 = 10.76 ft^2
= 1,000,000 mm^2

Pressure

inch of water, in WG = 0.0361 lb/in^2 = 5.196 lb/ft^2

pounds/square inch, lb/in^2 = 27.7 in WG = 2.036 in Hg
= 0.0703 kg/m^2 = 0.6896 newton/cm^2
= 6894.757 Pa
= 6894.757 newton/m^2

pascal, Pa = newton/m^2 = 0.00014504 lb/in^2

atmosphere, atm = 14.696 lb/in^2 absolute = 101.325 kPa
= 1.01325 bars
= 760 mmHg
= 29.92 in Hg

Density

pound/ft^3, lb/ft^3 = 16.019 kg/m^3

Glossary

aerosol Small particles, solid or liquid, suspended in air. The particle size may vary from 100 μ down to 0.01 μ or less, which can be dust, fog, or smoke.

air, standard Dry air at 70°F and 29.92 in Hg barometer pressure; air density is 0.075 lb/ft^3 and specific heat is 0.24 Btu/lb-ft.

airstream Airflow in a process duct, vessel, collector, or fan. This is usually the captured airflow with dust or vapors in it before passing through a collector, or the cleaned airstream after the collector.

atmosphere Air pressure at standard conditions, 14.696 lb/in^2, 29.92 in Hg (760 mmHg), 407.52 in WG (10,340 mm WG).

blast gate Sliding gate valve used to close off or partially open a duct to airflow; often used for balancing the airflow in branches or pickup from hoods in a dust collection system.

brake horsepower (bhp) Actual delivered or required horsepower, as different from rated horsepower. Output of a motor required to drive a fan, feeder, conveyor, or other equipment. This must include any losses in a V-belt or chain drive.

branch duct Smaller or feeder ducts that contribute airflow to a main duct or plenum. These branches need to be sized for the expected air volume and velocity required, with some allowance for balancing in the field as requirements change.

Btu British thermal unit, 1 Btu/h = 0.293 W.

bulk density Actual density of a dust or product as handled. This will vary from a fluidized lighter density, up to a settled density after enough time to allow air to escape and thus reach a stable maximum density.

°C Degree Celsius in the metric or SI system of units.

coefficient of entry C_e Ratio of actual to theoretical flow entering a hood or inlet caused by the static pressure in the hood or duct inlet and affected by the design configuration of the hood or inlet.

cubic foot Unit of volume measuring 1 ft long by 1 ft wide by 1 ft high, or its equivalent.

d Density factor of air at different conditions, d being 1 at standard conditions.

density Ratio of the mass of a substance to the volume of the specimen.

direct drive Motor shaft mechanically coupled to the driven shaft so that both turn at the same speed, usually with negligible power loss.

drive losses Losses in slippage and friction in a drive transmission, usually about 7 percent in a V-belt drive.

duct Enclosed tube or passage for constraining air or gas flow within the enclosure until it is delivered to a desired destination, such as a collector or fan, or exhausted to atmosphere.

dust Solid particles, usually small and light enough to be conveyed or carried in an airstream.

dust collector Separator of dust from an air or gas stream.

entry loss h_e Loss in pressure (in WG) caused by air flowing into a hood or duct inlet, due to the vena contracta of the airstream and shape constrictions in the inlet or hood.

°F Degree Fahrenheit.

ft/min Feet per minute, a measure of velocity.

ft/min (surface) Surface feet per minute, velocity over a surface.

ft/s Feet per second, a measure of velocity.

ft^2 Square feet, a measure of area.

ft^3/min Cubic feet of air flowing past a fixed location in 1 min.

ft^3/min (actual) Cubic feet per minute, at actual conditions.

ft^3/min (standard) Cubic feet per minute, at standard conditions of temperature and humidity.

fumes Often noxious suspension of minutely small solid particles in a gas or air, formed by the condensation of vapors with solid particles, or any smoke, vapor, or gas especially when irritating or offensive.

g Gram of weight, 453.6 g = 1 pound (lb) = 16 oz = 15.432 gr.

gases Formless fluids which tend to occupy uniformly an entire space at temperatures and pressures which do not change the substance into either a solid or liquid.

gr Grain of dust, an extremely fine weight measure, where 7000 gr = 1 lb, 1 gr = 0.0648 g = 0.0001429 lb, 1 gr/ft^3 = 2288.1 mg/m^3, and 0.01 gr/ft^3 = 22.88 mg/m^3.

gravity, specific, sp. gr. Ratio of the mass of a unit volume of a substance to the mass of the same volume of a standard substance at a standard temperature. Water at 39.2°F is the standard substance usually referred to. For gases, the mass of a unit volume of a gas divided by the mass of dry air at the same temperature and pressure as the gas.

h Hour, a 60-min time period.

h_e (See entry loss.)

HEPA High-efficiency particulate air filter. Usually about 99.95 percent efficient at collection of dust down to a 0.3-μ size.

hood Particular design of inlet used to capture dust- or fume-contaminated air or gas and conduct it into the collection and exhaust duct system.

hp Horsepower, a measure of motive or electric power.

humidity, absolute Weight of water vapor per unit volume in pounds per cubic foot or grams per cubic centimeter.

humidity, relative Ratio of the actual partial pressure of the water vapor in a space to the saturation pressure of pure water at the same temperature.

inch of water, in WG Unit of pressure equal to the pressure exerted by a column of liquid water 1 in high at a standard temperature.

in^2 Square inch, a measure of area.

lb Pound, a measure of weight.

lower explosive limit, LEL Lower limit of flammability or explosibility of a gas or vapor at ordinary ambient temperature, expressed in percent of the gas or vapor in air by volume. This limit is assumed to be constant for temperatures up to 250°F (121°C). Above these temperatures, it should be decreased by a factor of 0.7 since explosibility increases with higher temperatures.

magnahelic Pressure measuring instrument that senses the difference in air pressure exerted on opposite sides of a diaphragm, and by means of a magnetic linkage from the diaphragm to a pointer, indicates the pressure difference by the pointer motion. Usually the gauge reads pressure by inches water gauge for dust collection systems.

manometer Instrument for measuring air or gas pressure. It is essentially a U-tube partially filled with a liquid and so constructed that the amount of displacement of the liquid indicates the pressure exerted on the sensing end of the manometer tube.

ME Mechanical efficiency.

mg Milligram, or one-thousandth of a gram.

micron Unit of length measurement or size. It is one-thousandth of a millimeter, or one-millionth of a meter, or approximately 1/25,000 of an inch. It is identical to a micrometer.

min Minute of time.

minimum duct velocity Minimum airstream velocity required to move a specific type of dust particulate in the airstream, usually taken to be a conservative or safe conveying velocity that keeps the dust suspended in the airstream, but not higher than necessary for this suspension.

mist Small droplets of materials that are ordinarily liquid at normal temperature and pressure and are suspended in the air, like water in the form of droplets floating or falling in the atmosphere, or a suspension of a finely divided liquid in a gas, or a fine spray.

mm millimeter, or one-thousandth of a meter.

photohelic Differential pressure gauge and switch that incorporates two adjustable set point indicators and switches. Pressure sensing uses the same principle as the magnahelic gauge, and switching is accomplished by phototransistors that detect the gauge pointer when it reaches either set point of pressure level. Each set point is controlled by a knob on the front of the gauge. When either phototransistor switch is activated, it then actuates one of two internal relays, and contacts on both relays may be connected and used for control action circuits.

ppm Parts per million.

pressure, static, SP Pressure exerted in all directions by a fluid at rest. For a fluid in motion it is exerted in a direction normal or perpendicular to the direction of flow. It is usually expressed in inches water gauge for dust collection airstreams. SPa is actual SP at other than standard pressure. SPh is hood static pressure. SPs is SP at standard pressure.

pressure, total, TP Algebraic sum of the static and the velocity pressures.

pressure, vapor Pressure exerted by a vapor. The term vapor pressure is sometimes used synonymously with saturated vapor pressure.

pressure, velocity, VP Kinetic pressure measured in the direction of flow, which is necessary to cause the fluid to flow at a given velocity, usually measured by inches water gauge.

psi Pounds per square inch, a measure of pressure.

psig Pounds per square inch gauge, a measure of pressure difference from the surrounding atmosphere.

Q Quantity of airflow, measured in ft^3/min.

°R Degree Rankin, of temperature.

RAL Rotary air lock, a type of volumetric feeder or discharge equipment that has six or eight radial vanes that make up the rotor; the rotor is rotated inside a housing with close clearance around the rotor, and there is an upper inlet opening and a lower outlet opening.

RH Relative humidity; ratio of actual partial pressure of water vapor to the saturation pressure.

return air Cleaned or filtered airstream exhausting from a dust collector that is returned to a process, or to a plant area, as replacement or make-up air supply.

rpm Revolutions per minute, a measurement of rotating speed.

slot velocity Linear velocity or rate of flow through a slot opening, usually measured in feet per minute.

smoke Air suspension of particles, or aerosol, usually solids less than a μ in length or effective diameter from a combustion or sublimation source.

sp. gr. Specific gravity.

STEL Short-term exposure limit.

STP Standard temperature and pressure.

temperature, dry bulb Temperature level indicated by a thermometer without any effect of moisture on the reading.

temperature, effective An index that combines the effect of temperature, humidity, and air movement velocity on the sensation felt by the human body.

temperature, wet bulb Thermodynamic wet bulb temperature at which liquid or solid water can by evaporation bring the air to saturation adiabatically at the same temperature. Wet bulb temperature, without any qualifications, is the temperature indicated by a psychrometer, such as a sling psychrometer.

threshold limit values Values for airborne toxic material concentrations that people may be exposed to for 8 h/day, usually.

transport velocity Conveying velocity for dust in a duct or pneumatic conveyor pipe, usually the same or more than the minimum duct velocity in dust collection systems.

ULPA Ultralow particulate arrestor, particulate filter or arrestor for very fine size dust.

vapor Gaseous form of substances. Vapors diffuse. They can be changed to a solid or liquid state by an increase in pressure or decrease in temperature.

V_c Can velocity, or velocity of the airstream while it passes between the filter elements in a dust collector *can* or enclosure.

velocity Flow rate of an airstream, or other movement, usually measured in feet per minute.

velocity pressure (See pressure, velocity.)

vena contracta Formed by the convergence of the airstream just after it enters a hood or duct inlet. The area of the airstream at the vena contracta usually ranges from 70 to 100 percent of the duct area. A loss of static pressure occurs in the formation of the vena contracta, so it is desirable to use efficient hood and inlet designs.

V_f Filtration velocity, or velocity of the airstream when it enters or passes through the filter media, often referred to as the air-to-cloth (*A/C*) ratio determined by the air flow *Q* in cubic feet per minute of the airstream divided by the area of the filter elements or cloth in square feet, which equals feet per minute, or metric eqvivalents.

volume Measure of space, expressed in cubic feet or cubic meters.

Index

ABOUT THE AUTHOR

Miles L. Croom has more than 43 years experience in industry in the design and application of dry bulk handling and dust collection systems. He retired in 1993 as engineering manager of the Staclean Diffuser Company, a manufacturer of filter dust collectors, and now is consulting with the DCAD Company in Lexington, N.C. Mr. Croom received a B.S. in mechanical engineering from Rice University. He has published several papers on bulk handling, the most recent on "Effective Selection of Filter Dust Collectors" in *Chemical Engineering* magazine, July 1993. In November 1993 he was named national chair of the American Institute of Chemical Engineers New Procedure Committee to develop a performance testing procedure for gas filtration. He is a member of the American Institute of Chemical Engineers, and a former longtime member of the American Society of Mechanical Engineers, and the Society of Manufacturing Engineers. In 1989 he received a "Pollution Prevention Challenge Grant" award from the governor of North Carolina.